Artificial Intelligence Technologies in Management and Engineering

Artificial Intelligence Technologies in Management and Engineering

Edited by

Carolina Machado
J. Paulo Davim

iSTE

WILEY

First published 2026 in Great Britain and the United States by ISTE Ltd and John Wiley & Sons, Inc.

ISTE Ltd
27-37 St George's Road
London SW19 4EU
UK

www.iste.co.uk

John Wiley & Sons, Inc.
111 River Street
Hoboken, NJ 07030
USA

www.wiley.com

Library of Congress Control Number: 2026935275

British Library Cataloguing-in-Publication Data
A CIP record for this book is available from the British Library
ISBN 978-1-83669-052-8

The manufacturer's authorized representative according to the EU General Product Safety Regulation is Wiley-VCH GmbH, Boschstr. 12, 69469 Weinheim, Germany, e-mail: Product_Safety@wiley.com.

Contents

Preface

Artificial Intelligence (AI) technologies have played a transformative role in various fields, including management and engineering. The advancement of machine learning algorithms, increasing computing power and the availability of large volumes of data have driven the adoption of these technologies, making them indispensable for process optimization and strategic decision-making.

AI has been applied in management to improve operational efficiency, predict market trends and automate repetitive processes. Tools such as predictive analytics, based on big data, allow companies to anticipate demands and adjust their strategies based on patterns identified in historical data. Chatbots and virtual assistants also play a crucial role in customer service, helping to reduce costs and increase consumer satisfaction. AI systems are also used to assess financial risks, optimize supply chains and personalize consumer experiences, making business operations more agile and efficient.

Another relevant aspect is the use of AI in people management. Advanced tools analyze resumes, identify talent and predict which employees are most likely to leave the company, enabling more strategic human resource management. Additionally, AI systems help train employees through adaptive learning platforms, which adjust content based on employees' performance and individual needs.

In engineering, AI has revolutionized the development and maintenance of infrastructure. In civil engineering, for example, computer vision algorithms are used to analyze satellite images and predict structural failures in bridges and buildings, preventing accidents. In mechanical and electrical engineering, AI is used to enhance product design, resulting in greater efficiency and reduced waste. Industrial automation, powered by artificial intelligence, enables factories to operate with high precision, reducing human error and increasing productivity.

Another area where AI has demonstrated a significant impact is predictive maintenance. Smart sensors and machine learning algorithms enable continuous monitoring of industrial equipment, identifying patterns that indicate potential failures. This allows corrective actions to be taken before a breakdown occurs, avoiding unexpected downtime and reducing maintenance costs. Furthermore, AI has been used in robotics, with autonomous systems applied to civil construction and production engineering, thereby enhancing the safety and efficiency of these processes.

Software engineering also benefits greatly from AI, with tools that automate testing, detect security vulnerabilities and help generate code more efficiently. This reduces development time and improves the quality of technological products, ensuring greater reliability and performance.

However, despite the numerous advantages, implementing AI in management and engineering also presents challenges. Ethical issues such as data privacy and the impact of automation on the job market need to be carefully considered. Furthermore, adapting companies and professionals to these new technologies requires investments in qualifications and technological infrastructure. Businesses and governments must work together to ensure a balanced transition, minimizing negative impacts and maximizing the benefits of AI.

In short, AI technologies are shaping the future of management and engineering, delivering significant advances in efficiency and innovation. With the continuous evolution of these tools, it is expected that new business models and smart solutions will continue to emerge, promoting a positive impact on various sectors of the economy and society. To achieve this, organizations must invest in research, development and professional training, ensuring that artificial intelligence is used ethically and sustainably to drive progress.

Conscious of this reality, this book, entitled *Artificial Intelligence Technologies in Management and Engineering*, provides relevant support to academics and researchers, as well as those operating in the management, manufacturing and industrial engineering fields, who need to deal with the challenges and potentials underlying the use of AI technologies. It seeks to contribute to the advancement of research in the field, provide practical insights for managers and engineers, and even serve as a basis for future technological innovations.

There are several contributions that the book provides, not only concerning the advancement of scientific and technical knowledge (exploring, for example, the impact of AI on the optimization of industrial and business processes); as well as in terms of improving efficiency and decision-making (suggesting strategies for implementing AI in business management and engineering); and the development of new tools and technologies (by presenting opportunities for innovation in AI applied to engineering and management and/or suggesting new models and algorithms to improve automated processes).

It is also important to highlight that another contribution provided by this book is based on the opportunity to contribute to an adequate reflection on the social and ethical impacts resulting from the use of these AI technologies.

Looking to provide a channel of communication to share and disseminate knowledge of new advances in AI technologies in management and engineering between academics/researchers, managers and engineers, *Artificial Intelligence Technologies in Management and Engineering*, organized into eleven chapters, begins by discussing in Chapter 1, *"From Algorithms to Applications: AI in Management and Engineering"*, while Chapter 2 highlights *"Generational Perspectives on AI (From Baby Boomers to Gen Z): Understanding, Perceived Usefulness, Motivation to Adopt, and Risk Perception"*. Chapter 3 focuses on the *"Smart Decisions: How AI is Transforming Everyday Management and Engineering Practices"*. Chapter 4 speaks about *"Integrating AI in Business Education: Bridging the Gap Between Disciplinary Knowledge and Business Performance"*. Chapter 5 explores *"Holistic Management Quo Vadis? Designing Management Dispositive and Metamorphic Possibilities in the age of AI"*. Chapter 6 deals with *"Mapping the Use of Generative AI in Spain's Advertising Sector: Current Trends and Future Challenges"*. Chapter 7 covers *"Emotional Nudging in the Rise of Affective Artificial Intelligence"* and Chapter 8 looks to discuss *"Agentic Artificial Intelligence in Marketing: Opportunities, Challenges, and Impact on Firm Performance"*. Chapter 9 discusses *"AI's Role in Marketing: Mapping the Evolution of Creativity"*, while Chapter 10 analyzes *"Unveiling Management Research's Thematic Evolution: An Unsupervised Machine Learning – Latent Dirichlet Allocation Perspective"*. Finally, we come to Chapter 11, which seeks to highlight *"The Use of Artificial Intelligence in Human Resource Management: Barriers, Opportunities, and Trends"*.

Artificial Intelligence Technologies in Management and Engineering can be used by academics, researchers, human resources managers, managers, engineers and other professionals in related matters with artificial intelligence technologies in management and engineering.

In summary, we can say that this book looks to:

– share knowledge and insights about AI technologies in management and engineering on a global scale;

– stay at the forefront of matters and the most recent research activities relating to AI technologies in management and engineering, particularly concerning innovative models, theories, strategies, processes and procedures;

– increase our knowledge and contribute to an ethical and responsible use of more effective AI technologies in management and engineering.

The Editors express their gratitude to ISTE-Wiley for this opportunity and their professional support. Finally, we would like to thank all chapter authors for their interest and willingness to contribute to this project.

Carolina MACHADO
Braga, Portugal

J. Paulo DAVIM
Aveiro, Portugal

March 2026

1

From Algorithms to Applications: AI in Management and Engineering

What began in the mid-20th century as expert systems and symbolic reasoning has evolved into advanced machine learning (ML), deep learning (DL) and intelligent agent technologies that have the potential to revolutionize decision-making and operational behaviors across industries. To technically inclined readers, understanding AI is not merely an inventory of algorithms but a conceptual appreciation that positions every AI type within its broader technological and organizational setting. The twin use of AI in management and engineering illustrates convergence and divergence in the uses of technologies. In the case of management, AI facilitates predictive analytics, automation of processes and intelligence augmentation to improve organizational productivity. In the case of engineering, it enables design automation, predictive maintenance and smart system integration, transforming the pillars of industry and infrastructure. This chapter places AI as the practical enabler and theoretical instrument, and it draws attention to a taxonomy of AI types and multiple functions AI has in achieving managerial and engineering creativity.

1.1. Introduction

Artificial intelligence (AI) application in engineering is facilitating enormous revolutions in various areas. AI assists in design and optimization by analyzing enormous datasets to predict failure and optimize system performance, particularly in civil engineering (Cubric 2020; Adeyeye and Akanbi 2024). For construction management, AI assists in controlling schedules, monitoring costs and quality, and observing safety, with the use of platforms such as the engineering machine learning automation platform (EMAP) (Choi et al. 2021; Parekh and Mitchell 2024; Xu and Guo 2025). AI is a key factor in product life cycle management in Industry 4.0 and smart manufacturing, maximizing design, manufacturing and service stages

Chapter written by Hamed TAHERDOOST and Mitra MADANCHIAN.

(Wang et al. 2021). In materials and manufacturing studies, AI and machine learning (ML) complement traditional modeling techniques, especially where mechanisms are complex or high in computational costs (Lin and Lu 2020). Aerospace engineering is boosted by AI and blockchain technology for supply chain management and fault identification (Abdulrahman et al. 2023).

AI-based applications in business management are transforming efficiency and decision-making. Intelligent decision support systems improve project management using data-driven models for fostering insight and acceleration (Wang 2023). AI-based risk management goes beyond statistical methods for minimizing errors and maximizing accuracy throughout project stages (Choi et al. 2021). Human resource management and international business management are also assisted by AI through automation and performance improvement (Madanchian and Taherdoost 2025). Apart from this, Kansei Engineering uses AI to infuse emotional and consumer-centric designing elements into the products, hence making them attractive (Nagamachi and Lokman 2016).

Problems of data quality, algorithm bias and explainability requirements for AI systems are serious concerns that need to be addressed to enable successful deployment (Naser 2021; Whang et al. 2023). With businesses still in pursuit of the potential of AI, there is likely to be greater focus on creating solid frameworks that allow for the effective and responsible application of AI technologies across industries.

The aim of this chapter is to provide an integral taxonomy-driven technical explanation of AI technology in the context of management and engineering. Rather than citing isolated case studies or merely mandated applications, the chapter conceptually maps AI categories – such as knowledge-based systems, machine learning, natural language processing, computer vision, and intelligent agents – to their roles in organizational management and in shaping engineering imagination. The goal is to give technical readers a uniform framework for understanding the variety of AI applications, the conceptual underpinnings of different approaches and the comparative ways in which these technologies enhance decision-making, process optimization and system design.

1.2. Foundations of artificial intelligence

The quest for defining intelligence, and particularly intelligence in the case of artificial systems, is at the forefront. Legg et al. (Legg and Hutter 2007) note that one of the problems inside AI is that there is no clear definition of intelligence. They suggest a mathematical formalization that aligns with machine intelligence and require a definition to be broad for various types of intelligence, such as artificial systems. This kind of formalization serves as a starting point for studying how AI is defined and evaluated. Context is one of the main concepts in AI, specifically how systems understand and react to information.

Sowa (1997) records the history of context in AI, where Peirce's work started it off. This study concludes that despite the colossal dispute regarding context, there is no single definition that can be used across all cases. Insertion of context in AI systems is required in a bid to enhance their interpretation and interaction capabilities. Marchesotti et al. (2004) also explain the importance of context in ambient-intelligence systems. They introduce structured approaches to contextual knowledge definition, which become critical while developing scalable and adaptable AI systems. This multi-layered modeling of context allows AI to efficiently handle heterogeneous data, thereby improving its decision-making capability. Wang (2019) critically examines the issues of how to define AI and recommends that a good working definition should meet ordinary usage, have well-defined boundaries and be suitable for rewarding research. He recommends that intelligence may be defined as "adaptation with too little knowledge and resources", a focus that underscores the adaptive character of intelligence in humans and machines. This definition aligns with the wider context of AI capabilities, namely under General-Purpose AI Systems (GPAIS).

This definition is in line with the overall discussion of AI capabilities, particularly under General-Purpose AI Systems (GPAIS). Triguero et al. (2024) explained GPAIS as systems that may perform a wide variety of tasks without being programmed on an individual case-by-case basis. They also present a taxonomy that distinguishes types of GPAIS according to their nature and limitations and also defines what is considered AI. Collective intelligence (CI) is the other cornerstone of AI. Szuba (2001) poses the question of CI being used to enhance AI systems, so its definition and measures of quality would be formalized. This piece raises questions about the trade-off between individual and collective intelligence, and that AI could possibly get improved by the implementation of both those advantages. The AI-enabling technological sub-systems also constitute a critical element in the definition of AI. Torkamani et al. (2017) refer to evidence-based treatments in modern medicine to demonstrate how AI technology facilitates the performance of high-definition assessments of human health. Table 1.1 summarizes the main capabilities of AI, specifying their purpose, methodology, area of research and examples.

Analyzing AI technologies requires us remembering their strengths and weaknesses. Big data, pattern identification, predictive analytics and decision-making are best handled by ML algorithms. ML algorithms usually require humongous amounts of data and are prone to being biased unless properly managed (Bannister and Connolly 2020).

Capability	Definition	Key methods	Current research focus	Applications	References
Perception	Ability to interpret and understand sensory data from the environment.	Sensor fusion, NLP, computer vision	Multimodal perception, cognitive computing models, world models	Autonomous vehicles, voice assistants, image recognition	Chen et al. (2019) and Fung et al. (2025)
Reasoning	Drawing logical conclusions from available information.	Abductive learning, logic programming, symbolic reasoning	Attention mechanisms, symbolic correction of learned facts	Medical diagnosis systems, intelligent tutoring systems	Dai et al. (2019) and Chen et al. (2020)
Decision-making	Selecting the best action based on perception and reasoning.	Decision algorithms, hybrid human–AI models	Mitigating cognitive biases, transparent AI systems, collaborative decision-making	Business analytics, automated trading	Shrestha et al. (2019), Shin (2021) and Rastogi et al. (2022)
Learning	Improving system performance through experience and data.	Machine learning, reinforcement learning, neural networks	Integration with reasoning and decision-making, adaptation to new situations	Robotics, recommendation systems, game AI	Gupta et al. (2022) and Putta et al. (2024)

Table 1.1. *A framework for understanding core capabilities of AI*

Conversely, NLP tools are strong in automating communication and processing human language but can be weak in context and ambiguity. Computer vision tools are also strong in automating manufacturing and quality inspection but might fail in changing environmental conditions (Nti et al. 2022).

The applicability of AI technologies varies significantly in management and engineering contexts. In management, AI can aid in decision-making activities, enhance customer relationship management and optimize supply chain processes. For example, AI-driven analytics can provide insights into consumer behavior, enabling organizations to modify their strategies (Wang et al. 2020). In contrast, in engineering, AI technologies tend to be used to facilitate design activities, product quality improvement and process optimization of manufacturing. Implementing AI in engineering processes can lead to innovations such as smart manufacturing systems and driverless vehicles, which rely on advanced algorithms for making real-time decisions (Tipaldi et al. 2020).

1.3. AI in management

Modern management relies increasingly on AI, which is revolutionizing the manner in which companies make decisions, streamline operations and react to emerging situations (Table 1.2). Theoretically, AI in management can be defined as a collection of theoretical models and algorithmic processes that enhance organizational acumen by imbuing machines with abilities such as sight, understanding and rationality within bureaucratic settings. With the developments in ML, DL and natural language processing, AI has been applied beyond using strict rules to adaptive process management, predictive models and self-execution decisions. To enhance strategic forecasting and risk management, managers can identify customer, accounting, and operational patterns from corresponding data using machine learning (ML) techniques, particularly supervised and unsupervised models (McAfee and Brynjolfsson 2017).

Although more complex, reinforcement learning is increasingly used in adaptive pricing and dynamic resource planning and is a reflection of the direct impact that AI techniques have on managerial decision-making in situations of high change (Sutton and Barto 1998). Supported by customer service chatbots, CRM sentiment analysis and management task-enabled smart virtual assistants, natural language processing has extended its impact on management by enhancing the user experience and the effectiveness of communication (Cambria and White 2014).

With these fundamental technologies, there are many opportunities across a broad set of functional areas with AI in management practice. Applications of AI in human resource management include its use in recruitment, where ML algorithms scan resumes, screen candidate fit and minimize selection prejudices (Delecraz et al. 2022; Chen 2023; Cai et al. 2024). AI also provides executives with evidence-based recommendations for enhancing corporate culture and retention through the use of workforce analytics, talent management and tracking of employee performance (Rožman et al. 2022; Farinha and de Fatima Pina 2025). AI-driven predictive analytics and recommendation systems in marketing management allow businesses to predict demand, create nimble advertising campaigns and customize customer experiences (Wedel and Kannan 2016).

Supply chain and operations management is another area where AI has been playing a revolutionary role. With predictive maintenance and dynamic scheduling, managers can reduce expenditures, optimize logistics networks and predict disruptions by using ML algorithms alongside real-time sensor data from IoT systems (Ivanov et al. 2021). Strategically, AI technologies give managers access to market simulation tools, competitive intelligence and scenario planning that enable decision-makers to try out theories and investigate alternative outcomes given different assumptions (Spangler 2002; Vudugula et al. 2023).

Domain	AI application	Contributions	References
Knowledge management	Development of knowledge bases and ontologies	Enhancing organizational knowledge sharing and decision-making	O'Leary (2002)
Supply chain management	Practical SCM problem-solving using AI; systematic review of AI techniques	Highlighting past successes and future opportunities in SCM optimization	Min (2010) and Toorajipour et al. (2021)
Agriculture and healthcare	AI in crop, weed and disease management; AI in health data management	Improving productivity in agriculture; transforming health information management	Eli-Chukwu (2019) and Stanfill and Marc (2019)
Disaster and telecom management	AI in disaster phases; zero-touch 5G network management	Supporting informed disaster decisions; exploring automated telecom operations	Benzaid and Taleb (2020) and Sun et al. (2020)
Organizational and emerging tech	Social/organizational AI impacts; AI in R&D decisions; AI/ML in fog and edge computing	Addressing adoption barriers; analyzing decision overreliance; managing dynamic computing environments	Cubric (2020), Keding and Meissner (2021) and Iftikhar et al. (2023)

Table 1.2. *Overview of AI applications and impacts across various management domains*

AI-powered advanced analytics platforms bring together global value chains, capital markets and customer interactions' data to enable risk management and corporate strategy, thereby making AI an organizational resilience driver (Davenport and Harris 2017).

By making it possible for businesses to track environmental performance, ensure compliance with regulations and align company operations with CSR objectives, AI enables governance and sustainability management (George et al. 2014).

The use of AI for management is not without challenges, even though these advantages are present. Algorithmic bias, data privacy and interpretability are still serious concerns, particularly when managerial choices affect customer trust, employees' lives and firm reputation (Mittelstadt et al. 2016).

AI hiring systems, for example, vow higher productivity but threaten to automatically copy historic biases in training data, posing ethical concerns for managers (Raghavan et al. 2020). Likewise, managers will be less accountable if they rely substantially on opaque DL systems for future planning because it may be problematic to justify decision results to stakeholders (Rudin 2019).

1.4. AI in engineering

The use of computational intelligence in engineering processes to leverage design creativity, optimize resource utilization and enhance reliability and safety is the fundamental concept of AI in engineering (Table 1.3).

Diagnostic assistance and trouble-shooting expert systems were the leading field of application of early AI to engineering, where rule-based reasoning encoded engineering knowledge as inflexible systems (Dym and Levitt 1991). However, with augmented application of ML and DL, engineering has seen a paradigm shift towards data-driven approaches capable of detecting sophisticated patterns in high-dimensional data provided by sensors, simulations and operating conditions without human intervention (Bishop and Nasrabadi 2006).

For predictive modeling applications such as material fatigue prediction, energy consumption forecasting and structural deterioration prediction, supervised learning is used most frequently (Hastie et al. 2009). In contrast, unsupervised learning is applied for outlier detection during manufacturing and monitoring of infrastructure to assist engineers in detecting errors or deviations without labeling beforehand (Chandola et al. 2009).

Optimization of performance in the automotive, energy and aerospace industries and predictive simulations are facilitated by digital twins, which simulate virtual models of physical systems and use AI mostly to revise models in real-time based on data from the real world (Tao et al. 2018).

Material science is now using AI-driven evolutionary algorithms and generative adversarial networks (GANs) to create novel alloys and composites with specific properties (Wei et al. 2019). Another important application of AI in engineering, especially in manufacturing and industrial processes, is predictive maintenance. Predictive maintenance through AI models can forecast the remaining useful life, identify early signs of equipment failure and suggest maintenance actions before expensive failures happen, via IoT sensor data analysis (Lee et al. 2015).

AI algorithms control dynamic processes, maximize energy consumption and level production loads on various equipment and lines in intelligent industrial environments, or Industry 4.0 (Schuh et al. 2017).

Robot arms, drones and self-driving vehicles can all accomplish intricate maneuvers with accuracy and adaptability with the assistance of reinforcement learning agents in robotics for adaptive motion planning (Kober et al. 2013).

As regulatory compliance and accountability become difficult to achieve with black-box models, the incorporation of AI into safety-critical systems highlights the mounting need for explainable and reliable AI in engineering (Samek et al. 2019).

Engineering area	AI methods commonly used	Primary purpose
Civil Engineering	Machine learning, computer vision, digital twins	Structural health monitoring, infrastructure design
Mechanical Engineering	Reinforcement learning, expert systems, robotics	Predictive maintenance, process automation
Electrical Engineering	Neural networks, intelligent agents, optimization AI	Power grid management, circuit design, signal analysis
Industrial Engineering	Unsupervised learning, multi-agent systems, NLP	Supply chain optimization, quality control
Chemical Engineering	Physics-informed neural networks, expert systems	Process control, reaction optimization
Aerospace Engineering	Reinforcement learning, computer vision, hybrid AI	Autonomous navigation, system diagnostics
Biomedical Engineering	Deep learning, computer vision, NLP	Medical imaging, prosthetics, personalized medicine

Table 1.3. *Use of AI in different engineering fields*

1.5. Comparative taxonomy of AI applications

Taxonomy categorizes the multifarious forms of AI and their theoretical, as well as, applied fields across various disciplines. They include knowledge-based systems, computer vision, natural language processing (NLP), supervised, unsupervised and reinforcement ML, intelligent agents and hybrid neuro-symbolic systems. Knowledge-based systems, which were earlier important to both industries, continue to offer value in rule-based setups where expert knowledge must be formalized for decision-making. Expert systems enable frameworks of compliance and decision support systems within management to allow systematic reasoning to take place in regulatory or financial audit procedures (Bailey et al. 1987; Amponsah and Ali 2023). Similar systems are also used in engineering for design verification, fault diagnosis and fault location, where technical problem-solving is maximized by codified rules (Dym and Levitt 1991).

The most frequent class within this taxonomy is ML, which offers predictive analytics, anomaly detection and optimization and acts as a link between engineering and management. Supervised ML algorithms are most commonly applied in demand forecasting, employee turnover prediction and financial risk analysis (Adeusi et al. 2024; Morelli et al. 2024) in management, whereas the same class of algorithms predicts wear of structures, energy consumption or component failure in advanced systems (Hastie et al. 2009) in engineering. Unsupervised learning has applications in both engineering and management. In engineering, it is used for sensor network fault detection, manufacturing quality control, anomaly detection, and structural health monitoring (Aldrich and Auret 2013; Eltouny et al. 2023), whereas in management it is applied to fraud detection, market segmentation and customer clustering (Hilas and Mastorocostas 2008; Harris and Austin 2024).

Comparative taxonomy discovers cross-domain common issues. Data quality and availability are symptoms of ubiquitous constraints, whether data silos in the organization constrain management analytics or noisy sensor readings constrain engineering models. Both disciplines share comparable ethical issues: in engineering, models which can be explained are required for safety-critical applications to provide accountability in the event of system failure, and in management, fairness of algorithmic hiring and transparency of financial decisions is an issue.

Large-scale computations in management and sophisticated engineering optimizations require substantial computational resources; therefore computational costs and scalability remain major concerns. The taxonomy also highlights complementary synergies: methods optimized for one domain often prove applicable in another.

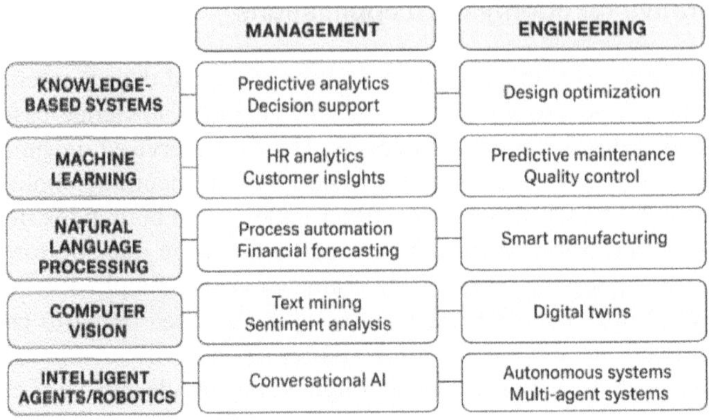

Figure 1.1. *Visualization of taxonomy*

For instance, anomaly detection techniques in engineering have influenced fraud detection systems used in financial management, and reinforcement learning designed for robotics has influenced adaptive decision-making in management. Figure 1.1 presents a taxonomy diagram that maps five main AI types to their respective applications in management and engineering.

1.6. Challenges and limitations

AI has the potential to revolutionize engineering and management, but there are several obstacles in its path. Interpretability and trust are blocked by the complexity and lack of transparency of most AI models, especially DL systems. The "black-box" nature of AI systems in engineering presents safety hazards if embodied in autonomous processes or critical infrastructure, and in management applications renders decision-making more challenging and undermines stakeholder trust. Performance is also limited by data availability and quality since AI needs large, representative datasets to generate valid predictions. Siloed data, biases and inconsistencies are common in most industrial and organizational environments, thus resulting in models that reinforce rather than fix underlying errors or inefficiencies.

Scalability and generalizability across diverse settings present a second challenge. The use of AI in management sometimes encounters difficulties when applied uniformly across organizations with different workflows, cultures or legal requirements. Engineering solutions crafted under controlled settings sometimes do not perform well when transferred to real-world contexts that are noisy, dynamic and uncertain. Also, availability is restricted by the expense and computational requirements of creating and maintaining upper-level AI systems, especially for small and medium businesses or low-budget engineering projects. As only large corporations with substantial financial resources can fully leverage AI advancements, obstacles such as these can enhance technological gaps.

The trajectory of AI adoption is also influenced by human, ethical and legal constraints. In managerial contexts, where AI-driven decisions directly impact stakeholders, customers and employees, privacy, accountability and fairness are of paramount concern. Engineering responsibility issues occur when autonomous systems break down, and nobody knows who to blame, the algorithm, the company or the coder. Additionally, if decision-makers over-depend on computational results, they lose human expertise and analytical thinking. Transparency, flexibility and inclusivity need to be taken into account in designing AI systems across all domains, supported by governance structures that balance innovation with organizational and societal safeguards, to ensure sustainable progress.

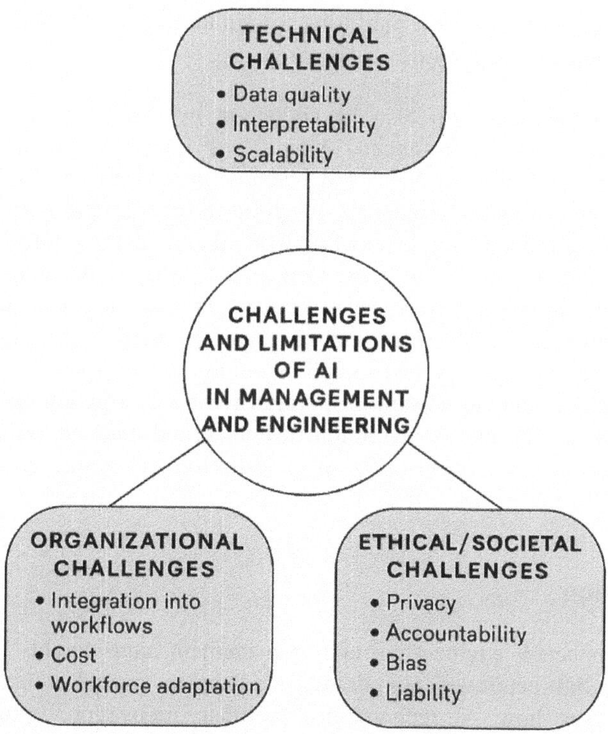

Figure 1.2. *Limitations of AI in management and engineering contexts*

1.7. Future directions

Transparency, accountability and interpretability will assume increasingly greater significance as businesses and industries rely increasingly on algorithmic decision-making. This change in management will foster enhanced stakeholder trust, especially in strategic planning, financial forecasting and human resource analytics where the rationale behind recommendations is as crucial as the result. Explainable AI will be critical in engineering for safety and reliability, especially in applications such as robots, autonomous systems and critical infrastructure monitoring. Future work will focus on hybrid models that integrate machine intelligence and human expertise to create systems that are not only able to predict, but also explain and evolve. Yet, another exciting direction is the convergence of AI with new digital worlds such as edge computing, digital twins and the Internet of Things (IoT). Real-time analysis of operations, customer interactions and market trends will be facilitated by this convergence in management, making businesses more responsive and agile. AI-powered digital twins in engineering will offer unprecedented system lifecycle insights with powerful platforms for simulation, design optimization and predictive maintenance capabilities.

Scalability challenges will also be addressed by the integration of AI with distributed computing environments, which will allow companies to deploy intelligent systems in decentralized and resource-limited environments without a loss of functionality. Such interconnections point to the fact that AI will ultimately function both as a fundamental layer of networked and adaptive infrastructures and as a decision support tool. The socio-technical aspects of AI adoption must be addressed going forward. Governance frameworks that ensure ethical behavior, diversity and fair access to AI technologies will be needed for engineering and management. As employees adjust to collaboration models where AI complements rather than replaces human capabilities, workforce transformation will continue to be a critical issue. The mutual evolution of human and machine work will be the subject of ongoing research, prioritizing multidisciplinary working, education and reskilling.

1.8. Conclusion

AI has impacted engineering and management considerably, becoming a homogenizing but heterogeneous field. The chapter created a taxonomy-based model to analyze how AI technologies facilitate innovation, optimization and decision-making through the classification of the technology into various types, including intelligent agents, ML, natural language processing, expert systems and computer vision. From a comparative perspective, engineering applies AI for design automation, real-time monitoring and integration of complex systems; while management targets knowledge improvement, predictive analysis of data and strategic counsel. Overall, these areas describe the manner in which AI is influencing technological as well as organizational settings.

However, the research also identified important limitations and challenges that should not be ignored. Major barriers to large-scale use still exist in areas of data quality, interpretability, scalability and responsibility. In environments where AI influences decisions that impact organizations and individuals, openness and trust are critical to management. When smart systems are integrated into autonomous technologies or physical infrastructures, safety, reliability and responsibility are non-negotiable issues in engineering. In addition to technological innovation, sound governance structures, multi-disciplinary collaboration and ongoing realigning between machine intelligence and human knowledge will be needed to handle these issues. The future of AI in management and engineering will depend on whether it can develop in an inclusive and responsible manner.

Technological innovation in explainable AI, digital twin and IoT integration, and socio-technical governance paradigms indicate actionable directions that strike a balance between responsibility and creativity. The future holds symbiotic models where technical precision and human judgment converge, with AI technologies taking on a facilitator role to support human decision-making, rather than becoming a substitute. This chapter's comparison and taxonomy offer technical readers a framework for understanding these developments, enabling AI to remain a powerful and enduring force for change in management and engineering.

1.9. References

Abdulrahman, Y., Arnautović, E., Parezanović, V., Svetinovic, D. (2023). AI and blockchain synergy in aerospace engineering: An impact survey on operational efficiency and technological challenges. *IEEE Access*, 11, 87790–87804.

Adeusi, K.B., Amajuoyi, P., Benjami, L.B. (2024). Utilizing machine learning to predict employee turnover in high-stress sectors. *International Journal of Management & Entrepreneurship Research*, 6(5), 1702–1732.

Adeyeye, O.J. and Akanbi, I. (2024). Artificial intelligence for systems engineering complexity: A review on the use of AI and machine learning algorithms. *Computer Science & IT Research Journal*, 5(4), 787–808.

Aldrich, C. and Auret, L. (2013). *Unsupervised Process Monitoring and Fault Diagnosis with Machine Learning Methods*. Springer, London.

Amponsah, I.A. and Ali, M.O. (2023). Examining the influence of expert systems and decision support systems on auditing. *Albukhary Social Business Journal*, 5, 47–61.

Bailey Jr, A.D., Meservy, R.D., Duke, G.L., Johnson, P.E., Thompson, W. (1987). Auditing, artificial intelligence and expert systems. In *Decision Support Systems: Theory and Application*. Springer, Berlin, Heidelberg, 265–319.

Bannister, F. and Connolly, R. (2020). Administration by algorithm: A risk management framework. *Information Polity*, 25(4), 471–490.

Benzaid, C. and Taleb, T. (2020). AI-driven zero touch network and service management in 5G and beyond: Challenges and research directions. *IEEE Network*, 34(2), 186–194.

Bishop, C.M. and Nasrabadi, N.M. (2006). *Pattern Recognition and Machine Learning*. Springer, New York.

Cai, F., Zhang, J., Zhang, L. (2024). The impact of artificial intelligence replacing humans in making human resource management decisions on fairness: A case of resume screening. *Sustainability*, 16(9), 3840.

Cambria, E. and White, B. (2014). Jumping NLP curves: A review of natural language processing research. *IEEE Computational Intelligence Magazine*, 9(2), 48–57.

Chandola, V., Banerjee, A., Kumar, V. (2009). Anomaly detection: A survey. *ACM Computing Surveys (CSUR)*, 41(3), 1–58.

Chen, Z. (2023). Collaboration among recruiters and artificial intelligence: Removing human prejudices in employment. *Cognition, Technology & Work*, 25(1), 135–149.

Chen, S., Jian, Z., Huang, Y., Chen, Y., Zhou, Z., Zheng, N. (2019). Autonomous driving: Cognitive construction and situation understanding. *Science China Information Sciences*, 62(8), 81101.

Chen, S., Jiang, M., Yang, J., Zhao, Q. (2020). Air: Attention with reasoning capability. In *European Conference on Computer Vision*. Springer.

Choi, S.-W., Lee, E.-B., Kim, J.-H. (2021). The engineering machine-learning automation platform (emap): A big-data-driven AI tool for contractors' sustainable management solutions for plant projects. *Sustainability*, 13(18), 10384.

Cubric, M. (2020). Drivers, barriers and social considerations for AI adoption in business and management: A tertiary study. *Technology in Society*, 62, 101257.

Dai, W.Z., Xu, Q., Yu, Y., Zhou, Z.H. (2019). Bridging machine learning and logical reasoning by abductive learning. In *Advances in Neural Information Processing Systems 32 (NeurIPS 2019)*.

Davenport, T. and Harris, J. (2017). *Competing on Analytics: Updated, with a New Introduction: The New Science of Winning*. Harvard Business Press, Brighton, MA.

Delecraz, S., Eltarr, L., Becuwe, M., Bouxin, H., Boutin, N., Oullier, O. (2022). Responsible artificial intelligence in human resources technology: An innovative inclusive and fair by design matching algorithm for job recruitment purposes. *Journal of Responsible Technology*, 11, 100041.

Dym, C.L. and Levitt, R.E. (1991). *Knowledge-based Systems in Engineering*. McGraw-Hill Book Company, New York.

Eli-Chukwu, N.C. (2019). Applications of artificial intelligence in agriculture: A review. *Engineering, Technology & Applied Science Research*, 9(4), 4377–4383.

Eltouny, K., Gomaa, M., Liang, X. (2023). Unsupervised learning methods for data-driven vibration-based structural health monitoring: A review. *Sensors*, 23(6), 3290.

Farinha, J. and de Fatima Pina, M. (2025). The use of artificial intelligence and machine learning in strategies for talent retention and leadership development. In *Evolving Strategies for Organizational Management and Performance Evaluation*, Marcão, R. and Santos, V. (eds). IGI Global, Hershey, PA, 627–644.

Fung, P., Bachrach, Y., Celikyilmaz, A., Chaudhuri, K., Chen, D., Chung, W., Dupoux, E., Gong, H., Jégou, H., Lazaric, A. et al. (2025). Embodied AI agents: Modeling the world. arXiv preprint, arXiv:2506.22355.

George, G., Haas, M.R., Pentland, A. (2014). *Big Data and Management*. Academy of Management, Briarcliff Manor.

Gupta, S., Modgil, S., Bhattacharyya, S., Bose, I. (2022). Artificial intelligence for decision support systems in the field of operations research: Review and future scope of research. *Annals of Operations Research*, 308(1), 215–274.

Harris, F. and Austin, V. (2024). Comparative study of supervised and unsupervised machine learning approaches in banking applications. SSRN 5394181.

Hastie, T., Tibshirani, R., Friedman, J. (2009). *The Elements of Statistical Learning*. Springer, New York.

Hilas, C.S. and Mastorocostas, P.A. (2008). An application of supervised and unsupervised learning approaches to telecommunications fraud detection. *Knowledge-Based Systems*, 21(7), 721–726.

Iftikhar, S., Singh Gill, S., Song, C., Xu, M., Sadegh Aslanpour, M., Toosi, A.N., Du, J., Wu, H., Ghosh, S., Chowdhury, D. et al. (2023). AI-based fog and edge computing: A systematic review, taxonomy and future directions. *Internet of Things*, 21, 100674.

Ivanov, D., Dolgui, A., Sokolov, B. (2021). The impact of digital technologies on supply chain resilience. *Computers & Industrial Engineering*, 151, 107021.

Keding, C. and Meissner, P. (2021). Managerial overreliance on AI-augmented decision-making processes: How the use of AI-based advisory systems shapes choice behavior in R&D investment decisions. *Technological Forecasting and Social Change*, 171, 120970.

Kober, J., Bagnell, J.A., Peters, J. (2013). Reinforcement learning in robotics: A survey. *The International Journal of Robotics Research*, 32(11), 1238–1274.

Lee, J., Bagheri, B., Kao, H.-A. (2015). A cyber-physical systems architecture for industry 4.0-based manufacturing systems. *Manufacturing Letters*, 3, 18–23.

Legg, S. and Hutter, M. (2007). Universal intelligence: A definition of machine intelligence. *Minds and Machines*, 17(4), 391–444.

Lin, G. and Lu, N. (2020). Integrating artificial intelligence, theory, modeling and experiments–perspectives, challenges, and opportunities in materials and manufacturing. *ES Materials & Manufacturing*, 9(2), 1–2.

Madanchian, M. and Taherdoost, H. (2025). Barriers and enablers of AI adoption in human resource management: A critical analysis of organizational and technological factors. *Information*, 16(1), 51.

Marchesotti, L., Piva, S., Regazzoni, C. (2004). Structured context-analysis techniques in biologically inspired ambient-intelligence systems. *IEEE Transactions on Systems, Man, and Cybernetics-Part A: Systems and Humans*, 35(1), 106–120.

McAfee, A. and Brynjolfsson, E. (2017). *Machine, Platform, Crowd: Harnessing Our Digital Future*. WW Norton & Company, New York.

Min, H. (2010). Artificial intelligence in supply chain management: Theory and applications. *International Journal of Logistics: Research and Applications*, 13(1), 13–39.

Mittelstadt, B.D., Allo, P., Taddeo, M., Wachter, S., Floridi, L. (2016). The ethics of algorithms: Mapping the debate. *Big Data & Society*, 3(2), 2053951716679679.

Morelli, C., Fusai, G., Zenti, R. (2024). Who and why will leave me? Utilizing machine learning-based models to anticipate and manage employee turnover. Working Paper.

Nagamachi, M. and Lokman, A.M. (2016). *Innovations of Kansei Engineering*. CRC Press, Boca Raton, FL.

Naser, M. (2021). An engineer's guide to eXplainable Artificial Intelligence and Interpretable Machine Learning: Navigating causality, forced goodness, and the false perception of inference. *Automation in Construction*, 129, 103821.

Nti, I.K., Adekoya, A.F., Weyori, B.A., Nyarko-Boateng, O. (2022). Applications of artificial intelligence in engineering and manufacturing: A systematic review. *Journal of Intelligent Manufacturing*, 33(6), 1581–1601.

O'Leary, D.E. (2002). Using AI in knowledge management: Knowledge bases and ontologies. *IEEE Intelligent Systems and Their Applications*, 13(3), 34–39.

Parekh, R. and Mitchell, O. (2024). Progress and obstacles in the use of artificial intelligence in civil engineering: An in-depth review. *International Journal of Science and Research Archive*, 13(1), 1059–1080.

Putta, P. et al. (2024). Agent q: Advanced reasoning and learning for autonomous AI agents. arXiv preprint, arXiv:2408.07199.

Raghavan, M., Barocas, S., Kleinberg, J., Levy, K. (2020). Mitigating bias in algorithmic hiring: Evaluating claims and practices. In *Proceedings of the 2020 Conference on Fairness, Accountability, and Transparency*.

Rastogi, C., Zhang, Y., Wei, D., Varshney, K.R., Dhurandhar, A., Tomsett, R. (2022). Deciding fast and slow: The role of cognitive biases in AI-assisted decision-making. *Proceedings of the ACM on Human-Computer Interaction*, 6(CSCW1), 1–22.

Rožman, M., Oreški, D., Tominc, P. (2022). Integrating artificial intelligence into a talent management model to increase the work engagement and performance of enterprises. *Frontiers in Psychology*, 13, 1014434.

Rudin, C. (2019). Stop explaining black box machine learning models for high stakes decisions and use interpretable models instead. *Nature Machine Intelligence*, 1(5), 206–215.

Samek, W., Montavon, G., Vedaldi, A., Hansen, L.K., Müller, K.R. (2019). *Explainable AI: Interpreting, Explaining and Visualizing Deep Learning*. Springer, Cham.

Schuh, G., Anderl, R., Gausemeier, J., Ten Hompel, M., Wahlster, W. (2017). Industrie 4.0 maturity index. *Managing the Digital Transformation of Companies*, 61, 1–64.

Shin, D. (2021). The effects of explainability and causability on perception, trust, and acceptance: Implications for explainable AI. *International Journal of Human–Computer Studies*, 146, 102551.

Shrestha, Y.R., Ben-Menahem, S.M., Von Krogh, G. (2019). Organizational decision-making structures in the age of artificial intelligence. *California Management Review*, 61(4), 66–83.

Sowa, J.F. (1997). Peircean foundations for a theory of context. In *International Conference on Conceptual Structures*. Springer, Berlin, Heidelberg.

Spangler, W.E. (2002). The role of artificial intelligence in understanding the strategic decision-making process. *IEEE Transactions on Knowledge and Data Engineering*, 3(2), 149–159.

Stanfill, M.H. and Marc, D.T. (2019). Health information management: Implications of artificial intelligence on healthcare data and information management. *Yearbook of Medical Informatics*, 28(01), 056–064.

Sun, W., Bocchini, P., Davison, B.D. (2020). Applications of artificial intelligence for disaster management. *Natural Hazards*, 103(3), 2631–2689.

Sutton, R.S. and Barto, A.G. (1998). *Reinforcement Learning: An Introduction*. MIT Press, Cambridge.

Szuba, T.M. (2001). *Computational Collective Intelligence*. Wiley, Hoboken, NJ.

Tao, F., Zhang, H., Liu, A., Nee, A.Y. (2018). Digital twin in industry: State-of-the-art. *IEEE Transactions on Industrial Informatics*, 15(4), 2405–2415.

Tipaldi, M., Feruglio, L., Denis, P., D'Angelo, G. (2020). On applying AI-driven flight data analysis for operational spacecraft model-based diagnostics. *Annual Reviews in Control*, 49, 197–211.

Toorajipour, R., Sohrabpour, V., Nazarpour, A., Oghazi, P., Fischl, M. (2021). Artificial intelligence in supply chain management: A systematic literature review. *Journal of Business Research*, 122, 502–517.

Torkamani, A., Andersen, K.G., Steinhubl, S.R., Topol, E.J. (2017). High-definition medicine. *Cell*, 170(5), 828–843.

Triguero, I., Molina, D., Poyatos, J., Del Ser, J., Herrera, F. (2024). General purpose artificial intelligence systems (GPAIS): Properties, definition, taxonomy, societal implications and responsible governance. *Information Fusion*, 103, 102135.

Vudugula, S., Chebrolu, S.K., Bhuiyan, M., Rozony, F.Z. (2023). Integrating artificial intelligence in strategic business decision-making: A systematic review of predictive models. *International Journal of Scientific Interdisciplinary Research*, 4(1), 01–26.

Wang, P. (2019). On defining artificial intelligence. *Journal of Artificial General Intelligence*, 10(2), 1–37.

Wang, J. (2023). Intelligent decision support system for building project management based on artificial intelligence. In *Journal of Physics: Conference Series*. IOP Publishing, Bristol.

Wang, Z., He, B., Yang, Y., Shen, C., Peña-Mora, F. (2020). Building a next generation AI platform for AEC: A review and research challenges. In *Proc. 37th CIB W78 Information Technology for Construction Conference (CIB W78)*, São Paulo.

Wang, L., Liu, Z., Liu, A., Tao, F. (2021). Artificial intelligence in product lifecycle management. *The International Journal of Advanced Manufacturing Technology*, 114(3), 771–796.

Wedel, M. and Kannan, P. (2016). Marketing analytics for data-rich environments. *Journal of Marketing*, 80(6), 97–121.

Wei, J., Chu, X., Sun, X.-Y., Xu, K., Deng, H.-X., Chen, J., Wei, Z., Lei, M. (2019). Machine learning in materials science. *InfoMat*, 1(3), 338–358.

Whang, S.E., Roh, Y., Song, H., Lee, J.G. (2023). Data collection and quality challenges in deep learning: A data-centric ai perspective. *The VLDB Journal*, 32(4), 791–813.

Xu, G. and Guo, T. (2025). Advances in AI-powered civil engineering throughout the entire lifecycle. *Advances in Structural Engineering*, 28(9), 1515–1541.

2

Generational Perspectives on AI (From Baby Boomers to Gen Z): Understanding, Perceived Usefulness, Motivation to Adopt and Risk Perception

The technological advance in artificial intelligence (AI) has transformed it into a burgeoning trend, with tremendous popularity in almost all aspects of human life. The surge of interest in AI has also fueled extensive research on the topic, aiming to understand a user's motivations for using this technology, its perceived usefulness, as well as the perceived risks associated with it. This chapter presents an exploratory study that aims to contribute to the growing literature on AI by providing an understanding of these aspects across different generations, specifically baby boomers, Gen X, millennials and centennials. Findings from this research show that AI is broadly viewed as useful and functional across cohorts, but the benefit depends on how individuals integrate specific tools into their routines. While usefulness and functionality are relatively consistent, the gratifications sought and the risks perceived seem to vary by generation, underscoring the need to tailor AI communication, design and policy interventions to different generations.

2.1. Introduction

Artificial intelligence (AI) has emerged as a transformative technology, reshaping how individuals interact with the world across various domains, from communication and healthcare to entertainment and finance. Over the last decade, AI has evolved from being primarily an experimental field to becoming an increasingly integral part of everyday life, with important implications for economic

Chapter written by Flor MORTON, Teresa TREVIÑO-BENAVIDES, Daniel Javier de la Garza MONTEMAYOR and Ana Valdés LOYOLA.

productivity, social interaction and organizational practices. According to the Stanford HAI AI Index Report of 2025, the use of AI in organizations has grown from 55% in 2023 to 78% in 2024 [STA 25]. Similarly, recent surveys suggest that between 20% and 40% of workers in the US already use AI tools in their daily tasks [CRA 25]. Adoption is also widespread in academia, with approximately 86% of students reporting the use of AI in their studies [CAM 25]. Together, these figures reflect a rapidly expanding adoption curve of AI. This transition involves not only technological advances but also conceptual, ethical and organizational challenges.

Despite its widespread adoption, perceptions of AI can vary significantly across generational lines, shaped by each generation's unique experiences, values and exposure to technology. Understanding these differences is crucial for tailoring AI systems to the diverse needs, motivations and expectations of users, while also anticipating the risks and barriers each group may perceive. Therefore, the purpose of this research is to compare perspectives on AI across different generations. Specifically, the study aims to explore the perceptions of usefulness on AI of baby boomers, Gen X, millennials and centennials; understand the motivations of each generation to use AI; and identify the perceived risks associated with AI among these generations.

This chapter is structured as follows: first, a literature review addressing user's perceived usefulness of AI, motivations for adopting this technology and perceived risks associated with AI is provided. Second, details of the data recollection method and participants' profiles are explained. Third, findings of the research are presented and discussed, highlighting generational similarities and differences. Finally, practical implications and future research avenues are explored.

2.2. Literature review

2.2.1. *Perceived usefulness of artificial intelligence*

The concept of perceived usefulness has been linked to the acceptance of technology since the definition provided by [DAV 89]. If a user considers that the technology is useful, then that perception can be a determining factor that leads to its frequent use. After this, it would be used in subsequent studies that would measure whether individuals and organizations were willing to accept new ways of working, motivated by the perceived usefulness of new technologies [ADA 92]. What changed in the following years was the evolution of technology, but the variables and relations that were studied remained consistent. From systems came Internet platforms, and then social networks. In the past few years, the rise of AI has gained attention from academics worldwide [DEL 23].

In the case of the investigation by [BIT 20], it was assumed that the perceived usefulness that users had about the performance of AI could be key in its acceptance. Researchers recommend strengthening user trust in AI, which can result in its further adoption by more people. But even with this reality, it is possible to anticipate that marketing strategies may change in the not-so-distant future. [SAD 22] investigated this and found empirical evidence that the perceived usefulness of AI could lead to the innovation of marketing strategies. There is evidence of this with the use of algorithms and big data in advertising [ADA 21].

When the use of AI is observed with an industrial focus, the perception of usefulness can be a predictive factor in the intention to use AI, according to the study by [KEL 23]. But the authors also agree that the information on the use of new technologies is currently in its preliminary phase. It is important to note that there is currently a surge in interest in the study of AI, and one of the issues that remains constant is precisely that perceived usefulness has been a variable that has consistently been evaluated to measure the intention to adopt the technology. AI has consistently shown to have an important relationship in several studies.

However, even when we are in a field marked by uncertainty, different disciplinary areas recognize the need to adopt AI. In the case of a study carried out with accounting students, it was also shown that perceived usefulness was directly linked to the adoption of AI [SUD 23]. The study by [KIM 21] makes a distinction between the two roles of AI. The study is a distinction between the functional and the social. In their results, evidence was found that the perceived usefulness is greater with the functional use of AI than with the social one.

2.2.2. *Uses and gratifications theory and AI use*

The uses and gratifications theory has long served as a foundational framework for understanding media consumption patterns and the motivations that the audience seeks. It effectively explains how individuals use media to achieve goals and gratifications [FIL 12]. It has its origins in the 1940s, but was reconstructed in the 1970s. The foundations of the theory were developed by Katz and colleagues when they proposed the idea that audiences are not passive recipients and users of media, but rather active users who select media to fulfill needs and obtain gratifications [KAT 74]. Further, this notion suggests that when studying audiences of media, it is important to understand what people do with media rather than what media does to people [SHI 11]. Research addressing the theory has identified several functions of media, with some variations depending on the type of media it studies. Research focusing on traditional media such as radio, television, cable TV, VCR remote controls and digital TV, agrees on three main functions, which are information seeking, diversion and maintenance of personal identity [JEN 02, TRE 15].

This theory has been used to study audiences under new contexts as new technology emerges. For example, there is significant literature on the motivations to use the Internet, and results suggest that there are a) process gratifications, such as the enjoyment in searching and surfing the web; b) content gratifications, such as the information and knowledge obtained; and c) social gratifications, which comes from chatting and interacting with friends and other users [STA 04]. Additionally, research on uses and gratifications theory has also explored why people use different Internet-related applications, such as social networking sites. In such research, "social interaction" is the most common and important emerging theme, as well as information seeking, entertainment, relaxation, communication, convenience, expression of opinion, information sharking, inspiration and maintaining knowledge of others [RAA 08, WHI 13, TRE 19, MOR 20].

In this context, and along with the advent of AI, there arises a novel intersection between the uses and gratifications and AI, as there is a new and complex interaction between users and this new technology. Recent research has studied why people engage and use voice-powered AI assistants and integrate them into their lives. Results suggest that people obtain two main gratifications by interacting with virtual assistants such as Alexa, which are those related to the functional tasks they help to do, and those related to enabling users to become better versions of themselves [JON 22]. Other research has suggested that some users may use such voice-powered assistants because of the enjoyment, social status and social influence [WAL 23].

Research on the topic has continued recently, as progressively more options for virtual assistants become available and part of people's lives. For example, [CHO 21] studied Amazon Alexa, Google Home, Apple Homepod, Samsung Bixby and Microsoft Cortana. This study considered a user's motivations and the interaction with their post-consumption attitudes, satisfaction and intentions to continue using the technology. They found that the motivations to use virtual assistants are interaction, personal identity, conformity, life efficiency and information. Further, they suggest that the utilitarian-related motivations (life efficiency and information) are significant antecedents of all variables and behavioral intentions, such as satisfaction, and intention to continue using the technology. In other words, users intend to use voice virtual assistants as long as they continue providing information and make their lives efficient [CHO 21].

When studying motivations to use AI, it is also relevant to understand the concerns that users perceive when engaging with such technology. [TUN 17] studied intelligent personal assistant usage motivations, addressing at the same time the privacy issues and concerns related to their use. They suggest that because social motivation is important and a determinant of usage, this also serves as a peer pressure mechanism to engage with a technology that they consciously know collects and uses their personal information in an unprecedented manner. The authors explain how participants feel trapped and enrolled in the need to use such technologies because the denial of using it would probably bring unfavorable consequences from a social perspective [TUN 17].

AI applications may cater to various user needs and gratifications. For instance, recommendation algorithms analyze user preferences to offer better and more personalized content, fulfilling users' informational and entertainment needs. Virtual assistants such as Alexa and Siri, provide users with convenience and assistance, satisfying their informational and interpersonal needs. However, AI encompasses different technologies, not only virtual assistants, that simulate human intelligence. Considering this, the convergence of uses and gratifications theory, with AI in its different forms and applications, presents novel avenues for research and practical applications. By applying this theory to AI contexts, insights can be obtained regarding user behaviors and preferences. In practice, practitioners can develop AI systems that effectively cater to user needs and gratifications.

2.2.3. *Perceived risk*

Perceived risk and trust are critical elements influencing consumer behavior, especially in the context of emerging technologies such as AI. A consumer's perceived risk refers to the anticipation of negative consequences associated with a consumption decision [SOH 24]. Trust, on the other hand, involves the expectation of positive outcomes in situations perceived as risky [ROH 23]. Individuals inherently seek to avoid adverse outcomes, leading them to adopt various approaches to mitigate perceived risks, such as avoiding situations they view as potentially harmful.

Perceived risk encompasses six dimensions: financial risk, social risk, functional risk, time risk, health risk and psychological risk. Financial perceived risk involves concerns over potential loss of money or that the cost of a product or service is not worth the value it provides [CHE 07]. Functional or performance risk reflects doubts about whether a product will perform as expected without failing [PAL 16]. Physical risk refers to the potential harm or safety issues associated with a product [CHE 07]. Social risk arises from the potential for a purchase decision or technology adoption to elicit negative responses from a consumer's social networks [NAS 15], affecting social status or acceptance among peers. Time risk is the concern about the time investment required for the adoption and use of the product [HIR 06], including the time spent researching, purchasing and possibly returning or dealing with issues related to the product. Psychological risk involves psychological discomfort such as anxiety or other negative emotions that may result from a consumption decision [DHO 01].

These dimensions of perceived risk influence consumer adoption of new technologies. High levels of perceived risk can slow adoption rates as consumers may hesitate to embrace new technologies due to concerns about financial loss, performance issues or security risks [HAS 21]. For instance, initial resistance to adopt AI-powered home security systems was driven by worries about data privacy and potential hacking.

Understanding and addressing perceived risks associated with AI is essential to fostering the adoption of these new technologies. Research by [HAS 21] suggests that challenges of trust, financial risk perceptions and concerns about data security – identified in e-commerce research – can also apply to AI adoption. Due to AI being a complex and emerging technology, consumers often lack understanding and trust, which hinders their willingness to adopt it.

Moreover, understanding the risk perceptions of AI across generational cohorts (centennial, millennials, Generation X and baby boomers) is relevant because each generational cohort has different experiences with technology, which may shape their attitudes, concerns and trust in AI differently. Baby boomers, who have witnessed the rise of the digital era later in life [VEN 17], may exhibit higher levels of skepticism and concern regarding privacy and data security. In contrast, millennials and centennials, who have grown up in a highly digital environment [THA 21], may be more accustomed to rapid technological advancements but still have concerns about transparency and ethical implications. Generation X, who experienced the pre-digital and digital eras [CAL 19], may hold more pragmatic views, while still facing concerns about the functional risks and potential societal impacts of AI.

2.3. Methodology

A qualitative research approach was adopted, with semi-structured interviews as the main recollection method. A total of 39 individuals participated in the study: eight baby boomers, 13 Gen X, ten millennials and eight centennials. Some of the interviews were conducted via Zoom, while others were conducted at the participant's home, with a minimum duration of 30 minutes. Participants were selected using a purposive sampling with the condition that they use at least one tool of AI for personal or professional purposes. Each interview session was transcribed; the transcripts were raw data for our qualitative analysis.

2.4. Findings

2.4.1. *Baby boomers*

2.4.1.1. *Baby boomers's understanding of AI*

The opinion of baby boomers regarding their understanding of AI can be summarized into four key themes: 1) although they have all heard about AI in previous times of their lives, they are now seeing the topic everywhere. They mentioned that it is prevalent in television and that friends, family and coworkers talk about this all of the time. They recognize the topic's current relevance and the importance of keeping up-to-date, and mention that it is now almost impossible not to know about it; 2) they view this new era of AI as science fiction becoming a reality, the representation of the phrase "the future is now". They see AI as a robot that could replace the need for human intelligence; 3) although some of the interviewees define AI in depth as a complex machine, system or algorithm that searches for an answer based on a web of information, others mentioned that they do not know exactly what it is, and they cannot really define it since it is a very complex topic for them; 4) when asked to provide some examples of AI, some of the baby boomers struggled to mention a specific example. They commented that although they know that AI is everywhere nowadays, they cannot provide a specific answer since they do not fully comprehend where it is. On the other hand, other baby boomers who understand the topic a little bit more, mentioned that they know that AI is used on advertisements; it is used to create fake Tik Tok videos, and in Chat GPT.

> Heard about it many years ago, but I didn't pay attention to it until recently (Maria Emma, baby boomer).

> It's been about a year since the first time I heard about it, but recently I've been hearing about it more and more (Rodrigo, baby boomer).

> It's like a person…but it's artificial, it answers everything you can ask of it (Lorena, baby boomer).

> A well-equipped software that is intertwined with a lot of information on the web of the world. And that's why the system itself is nourished and is being nourished automatically each time (Rodrigo, baby boomer).

> It's all so advanced. But I like to inform myself on the topic, to know what it's all about (Rosa Maria, baby boomer).

> I don't know how to explain it (Maria Emma, baby boomer).

2.4.1.2. *Baby boomers' motivations to use AI*

Four main motivations to use AI were found within the baby boomer segment: 1) some interviewees mentioned that their main motivation to use AI is to facilitate their daily tasks and activities, which helps them save time and get their tasks done faster. It simplifies their lives with direct benefits for themselves and their families; 2) others are motivated to use AIs mostly to help them in their professional lives and to solve complex world problems. For instance, one person mentioned that they will continue to do market research and to improve customer service in their company; 3) baby boomers also mentioned that they are motivated to use AI to easily learn new things. They view it as accessible knowledge that can help improve their lives; 4) some of the interviewees mentioned that they do not feel a particular motivation to use this technology, or they might use it inadvertently since they do not truly understand what it is. Although they see the benefits that AI can bring, they feel like they are "on their way out" and it is the younger people's turn to take advantage of this new technology.

> You no longer have to go and search for information. But I just have to say – I want to know how to cook an apple pie-, and it takes you through the steps (Rodrigo, baby boomer).
>
> I like to use it for everything, even to improve my sleep (Ricardo, baby boomer).
>
> It can help solve complex world problems, in engineering, design, it can even help reduce poverty (Paul, baby boomer).
>
> It gives us the opportunity to learn things, even though we're older, it gives us the opportunity to keep us updated (Rosa Maria, baby boomer).
>
> I use it to better understand my mom's disease (Lorena, baby boomer).
>
> I'm already on my way out, but for you, you still have a lot of room to use it (Ricardo, baby boomer).
>
> I may inadvertently use it. Since I have little knowledge of artificial intelligence (Carlos, baby boomer).

2.4.1.3. *Baby boomer's perceived usefulness of AI*

Four key themes regarding the perceived usefulness of AI were found in baby boomer interviews: 1) interviewees mentioned that AI is quite useful in optimization of tasks, to increase efficiency and to save resources (cost and time) within professional or business contexts. They emphasized its application to engineering, design, accounting, data collection and analysis, and most professional fields; 2) AI provides an easy and inexpensive access to important information, enriching their knowledge in a very accessible way. This can help in academic, professional and personal settings; 3) baby boomers think that AI can improve the quality of their personal lives. They perceive it as giving efficiency to their day-to-day tasks and even to their health by improving their eating, exercise and sleeping habits; 4) some baby boomers view AI as something that has a huge potential to solve complex global and humanitarian problems. For instance, they mentioned that AI can help develop further technological advancements that can help in various ways, such as developing treatment for complex diseases or solve complex global and humanitarian problems, such as poverty and other vulnerabilities.

> The main advantages I see are in the intellectual and research spheres, but not in manual labor and crafts. I do not think that artificial intelligence still surpasses that (Rodrigo, baby boomer).

> Access to important information immediately and puts you face-to-face with anyone, allowing you to verify or confront what you know and what other people tell you (Ricardo, baby boomer).

> In the professional field, it simplifies everything, everything flows immediately (Ricardo, baby boomer).

> It gives you a lot of data and statistics very quickly, including a roadmap for using that data (Paul, baby boomer).

> It represents a major breakthrough that can lead to plenty of other breakthroughs (Maria Emma, baby boomer).

> In the humanitarian sphere, it can provide you with material to solve complicated problems (Paul, baby boomer).

2.4.1.4. *Baby boomer's perceived risks of AI*

The main risks perceived by baby boomers can be summarized in four key themes: 1) some of them mentioned that they feel scared that the world is not prepared for AI and that adequate controls or regulations are currently missing. They mentioned that they feel concerned that a widespread use of AI could increase the frequency or intensity of terrorist attacks, fake news, manipulation or deception campaigns, extortion, plagiarism and even fraud. They additionally mentioned that to mitigate these risks, the values and principles of good use must be fostered throughout society and adequate regulations must be put in place as soon as possible to avoid misuses; 2) baby boomers mentioned that they are very concerned that AI will eventually lead to humans becoming less intelligent and eventually being replaced by robots, with job losses especially in manual labor. For them, this represents a large pushback on the advancement of humanity; 3) they fear that with the advancement of AI, the “virtual world” will become enmeshed with the “real world”. This will result in people becoming more isolated and disconnected from real life, representing a huge loss in socialization abilities and community formation. Additionally, they mentioned that they fear for the loss of privacy and feeling “watched” through AI, since we cannot truly know who can access your information; 4) another perceived risk by baby boomers is the fear of struggling to adapt to this new era of AI. They mentioned that throughout their lives, they have gone through multiple technological advancements, and they feel that now they might be “too old to adapt” to this new era. This is important because the rapid technological change can feel overwhelming, and their livelihood might be impacted by failing to adapt to the change.

> When I was a child there were no computers and then, computers were created and the Internet was created, and then products to vaccinate computers came out because viruses arrived. Then this, which was once unthinkable… these are things that can unleash chaos, let’s hope it will be controlled (Rodrigo, baby boomer).
>
> You become distrustful because you may be attacked (Maria Emma, baby boomer).
>
> Adaptation in people of my age is complicated, it is not easy to change your mindset (Maria Emma, baby boomer).
>
> We are not prepared (Ricardo, baby boomer).
>
> Values and principles of good use must be fostered (Paul, baby boomer).

2.4.2. *Generation X*

2.4.2.1. *Gen X’s understanding of AI*

Gen X participants perceive AI as encompassing machines, systems and algorithms that emulate human thinking, learning and communication to facilitate human activities. Some even mentioned that AI is science fiction materializing into reality; they reflect on how the first time they heard about AI and the capabilities that machines have today were once solely depicted in science fiction movies during their childhood and adolescence. It was interesting to find out that this generation also perceives any kind of automatization as AI; this was mentioned by participants and was also identified when participants provided examples of AI. For example, some participants mentioned:

> I consider artificial intelligence to be any system that ultimately replicates or exhibits certain processes parallel or similar to humans, such as calculations, reasoning, and is also capable of communicating in a way that is similar to that of humans (Arturo, Gen X).

> Artificial intelligence, as I have understood it, consists of algorithms developed within machines, across different devices, to emulate or imitate human behavior in various aspects, primarily aimed at facilitating human tasks (Enrique, Gen X).

> It's a very subjective topic, perhaps it's like the automatization of processes without the need or interference of humans (Héctor, Gen X).

> [About the first time that he heard about AI] I can go back to science fiction movies from when I was a teenager, where I saw some of them [AI and capabilities that machines have today], like a possible future, now it is a reality (Alfonso, Gen X).

When participants of this generation were asked to mention some examples of AI, they mentioned robots, chatbots (such as ChatGPT and chatbots that companies use for customer service), algorithms for social media (particularly those used to detect customer preferences and send content that could be interesting for them), virtual assistants (such as Alexa and Siri), mapping apps (e.g. Waze and Google Maps), smart appliances (some participants mentioned Smart TVs and Smart refrigerators as examples of AI), gadgets (e.g. smartphone, computers, smartwatch) and autonomous cars. As some participants mentioned:

> For example, when you call a company and a machine answers, it starts giving you options and asking you questions. Or when you need to contact technical support for something online, and a machine responds. I also know about something called ChatGPT… What else falls under artificial intelligence?…I've seen robots in restaurants collecting dirty items and taking them away. I guess that counts too (Patricia, Gen X).

> Social networks and other apps that detect your preferences to send content that may interest you (Alonso, Gen X).

2.4.2.2. *Gen X's motivations to use AI*

Regarding the reasons participants of Gen X use AI or would use AI (in case they have not used it), three motivations were identified. Most participants mentioned facilitation of daily tasks and activities (e.g. cleaning the house, making calls while driving, making grocery lists, among other daily activities) as a reason to use AI. For example, some participants mentioned:

> Well, for everyday life, for example, you're driving in the car and you need to make a call. Now, you just tell the car and say, "call dad" and then it [virtual assistant] asks you, "Do you want to call dad's cell or do you want to call dad's work?" So, you say, "Well, cell phone" and it connects a call without needing to lift a finger. So, that's a very clear example of something very quick, very comfortable, very practical, saving you time, being safe, helping you in your daily life (Juan, Gen X).

> As a housekeeper, AI can help me a lot in my activities. For example, even for making a grocery list, I can use Alexa every day, recording so I don't forget. Like, "Hey Alexa, can you write down this and this item for me, please?" (Marcela, Gen X).

Learning and information seeking were also found as motivations to use AI. Participants expressed an interest in leveraging AI technologies to expand their knowledge base and acquire new skills, as well as to improve their information acquisition for work-related aspects. Some respondents mentioned:

> Discovering new things would motivate me to use it because there is a personal satisfaction in learning and developing new skills ... We have a blog within our company where, like a newsletter, we update and upload information for our clients. And we have seen that ChatGPT helps us a lot to complement, to integrate more information that perhaps we didn't have so readily available (Héctor, Gen X).

> To be at the level we need to be in order to compete, right now, you know? The more skilled we are in handling all kinds of tools and electronic devices that provide us with more information and better results, the more efficient we become and we don't fall behind (Alonso, Gen X).

Curiosity and Fear of Missing Out (FOMO) were identified as additional motivations for using AI, as illustrated by the response of one participant. This individual shared their inclination to engage with AI firsthand, rather than hearing about it from others, indicating a fear of missing out on the experience or understanding of this particular technology:

> I used it for the first time simply to know what it was like, so I wouldn't hear about it from others, and to see what it was, how it worked (Alfonso, Gen X).

2.4.2.3. *Gen X's perceived usefulness of AI*

It was observed that Gen X participants recognize the ability of AI to improve their performance in numerous ways and across different domains. For instance, participants emphasized how AI enhances or has the potential to enhance performance through facilitating processes and activities, providing comfort, optimizing efficiency, saving time and costs, making informed decisions and reducing errors, and providing access to more information and precise data. For example, participants mentioned:

> You can complete tasks in less time, um, with greater precision. With it, I mean, you reduce errors (Marcela, Gen X).

> It could be useful for finding information, to relate data from different articles, to define a possible research question, things like that (Alfonso, Gen X).

Informants perceive AI as particularly useful in various domains, including daily activities (such as finding addresses, making calls, cleaning, making grocery lists, among others), training (for acquisition of new skills), communication (enabling global connectivity), health and medicine, research, business activities (such as segmentation, product development and production processes), scholar tasks, and for addressing environmental (e.g. global warming) and social issues (e.g. disability). For instance, respondents mentioned:

> [AI is useful] For making your everyday life simpler… for example, in the kitchen, you now have a smart refrigerator that informs you of what you need in your pantry (Edelmiro, Gen X).

> It [AI] is very useful for the development of different products…yes for production (Viviana, Gen X).

> AI can be very useful in commerce, where they classify you based on what you like to buy or where you like to travel, your average expenses, or which restaurants you go to (Marcela, Gen X).

> There are robots with algorithms used in the medical industry, in medicine, for certain operations, highly precise surgical procedures are robotized, and it has been proven that they even have greater precision and effectiveness than humans (Enrique, Gen X).

2.4.2.4. *Gen X's perceived risks of AI*

While Gen X generally trusts AI's utility in enhancing various aspects of life, participants also recognize certain risks associated with its use. Among the risks perceived by participants, the loss of cognitive and social skills stands out as an important concern as well as the effects in human health (e.g. obesity and psychological disorders) of a potential reduction in mobility and social interaction due to an increased reliance on AI. Additionally, there are concerns about the quality and reliability of information in an AI-driven society; this is due to a diminished ability to discern between true and false information. As mentioned by some participants:

> Human beings will become more passive and with reduced cognitive abilities (Rodrigo, Gen X).

> There might be false information, a person can trust the information provided by artificial intelligence even when it is not accurate (Rodrigo, Gen X).

> Humans might stop moving, leading to sedentary behavior. So, I see this as a significant disadvantage for humans in general. I remember the movie Wally, where at the end of the movie, he arrives in a world where humans are completely lazy, even more obese, more useless. And that's the disadvantage and fear we have of reaching that point (Enrique, Gen X).

> Human beings could become very comfortable in front of a screen or a smartphone and lose that warmth, that social interaction, that has always characterized us as human beings. We are social creatures, we live in society, and ultimately, we cannot be isolated. We see countries like Japan, where there are cases of great isolation among the Japanese, leading to problems of depression and anxiety, psychological issues in human beings, and that is a major disadvantage. We, as human beings, must be very careful not to become too excited about the advantages of artificial intelligence. We must not neglect this aspect, as it could backfire and lead to problems for our society and humanity (Enrique, Gen X).

Furthermore, there is a concern regarding the potential for widespread job displacement due to advancements in AI and automation. For example, a participant mentioned:

> It's concerning because it is [AI] taking away jobs from other people, I mean, with all the advantages AI or robotizing jobs brings for companies, many jobs are disappearing, so it's really a double-edged

> sword. For some, it is good because they are saving a lot of money, but for others, it affects them, so it depends on how you look at it (Lupita, Gen X).

Participants also expressed concerns about the loss of control and misuse of AI, which can lead to potential harm to humans. This encompasses biases embedded within social media algorithms to manipulate people and violations to human rights such as privacy infringements, espionage, identity theft and extortion. For example, some participants mention:

> Elon Musk has said that it [AI] can get out of control, which can be dangerous. He even said that human beings can be very limited compared to what we are achieving with artificial intelligence (Hector, Gen X).

> With a photo someone can put your voice, make it seem like your mouth said things and who knows what else.... well, it's a significant danger ... I think that's a very, very powerful weapon that can be used for harm (Patricia, Gen X).

> Well, I believe one of the disadvantages is the fact that artificial intelligence can replace a human, and the potential misuse of artificial intelligence to harm humans themselves...there might be risks associated with using artificial intelligence simply through the use of bots to drive trends on social media (Obed, Gen X).

2.4.3. *Millennials*

2.4.3.1. *Millennials' understanding of AI*

The in-depth interviews revealed that millennial users associate AI with machines, systems and algorithms designed to replicate human cognitive processes, particularly in terms of learning and decision-making. Overall, many participants demonstrated a basic understanding of AI as a technology capable of mimicking human intelligence. It can be said that millennial users predominantly perceive AI as a sophisticated technological tool used for processing vast amounts of information and performing both complex and everyday tasks efficiently. Some examples of tools or systems that participants relate to AI includes ChatGPT, virtual assistants such as Alexa and Siri, GPS, automated responses in various tools and even Google.

When talking to millennial users, it was clear that they perceive AI tools as a transformative force that drives innovation, disruption and revolution across various industries and sectors. The majority of respondents view AI as a technological advancement, enabling the development of solutions to everyday tasks and using time effectively. Furthermore, many participants expressed optimism about AI's potential to revolutionize traditional processes in various aspects of life. Most millennials interviewed are now employees or business owners so they see AI's potential in their professional activities. Additionally, participants recognize AI as tools designed to make daily tasks easier, enhance efficiency and optimize resource utilization. In particular, respondents perceive AI-powered tools as invaluable assets for managing time, increasing productivity and generating ideas.

Despite acknowledging AI's potential benefits, consumers also reflected on a paradoxical perception characterized by both trust and distrust in AI tools. While some participants express confidence in AI's ability to deliver accurate results and improve decision-making, others showed reservations regarding its reliability, transparency and implications.

2.4.3.2. *Millennials' perceived usefulness of AI*

Talking about the perceived usefulness of AI, all participants that use AI tools some way or another highlighted AI's role in simplifying processes and improving their overall productivity. Overall, evidence shows a positive attitude toward AI tools from this generation of users. They are open and willing to try the technology and enjoy the benefits it can provide. One participant sums up this idea with the following quote:

> For me, it's the dream come true, intelligence (Carlos, GenY).

Results show that an important number of participants actively use AI technologies to facilitate various processes and activities in their daily lives. Particularly, millennials reported that they use AI tools to a) obtain information and research about topics; b) to help them with business tasks; c) to program reminders; d) for school task completion; and e) to obtain ideas.

One of the most frequently mentioned actual uses of AI is to obtain information about any particular topic or question users may have. Most participants agreed on the fact that AI can be used to answer questions they encounter during their day, with a strong focus on the tasks they do on their jobs. As mentioned by the following participant:

> Any questions you may have or any opinions you would like to know, you can consult them and they (AI) will give you their point of view or provide you with their answer (Martin, GenY).

Other participants mentioned they use AI tools to program reminders, especially using virtual assistants for scheduling appointments and setting alarms. Furthermore, although Generation Y participants are mostly dedicated to their jobs and families, some participants are advancing in their professional studies, by pursuing other degrees. In such cases, AI tools help them with their academic activities, both to ask questions, generate ideas and develop essays. Finally, other more tech-savy users in this generation mentioned they use AI tools in a more advanced manner, such as predictive analytics tools for business forecasting and to elaborate planning and management systems for their company operations.

Consumers overwhelmingly acknowledge AI's potential to optimize efficiency, save time and reduce costs in both personal and professional contexts. Most respondents emphasized AI's ability to automate repetitive tasks, obtain answers and information, and allocate resources more effectively, leading to significant time and cost savings over time. Participants mentioned that this is a positive aspect of AI, and in fact, it can be perceived as an opportunity to transform things, tasks and processes. As one participant reflected:

> [AI] is an opportunity to make better the things that we have (Eugenio, GenY).

Convenience emerges as a primary motivator driving consumers' adoption of AI technologies, with many participants valuing AI-powered products and services for their ease of use and accessibility.

Users also value the fact that AIs have the ability to provide accurate and reliable information, for most topics and questions they may have. Overall, evidence suggest that most participants have a favorable and positive attitude toward AI tools. This positive attitude is strongly influenced by the time-saving advantage that millennial users perceive. Additionally, it is important to note that most Gen Y participants do not see AI tools as a replacement for humans, rather an enhancement for their activities, and therefore, perceived as convenient. For example, the following participant mentioned:

> [AI tools are used] to facilitate work for humans, so that they can focus on things that AI cannot do. AI can help us arrive at solutions faster (Juan Carlos, Gen Y).

2.4.3.3. *Motivations to use AI tools*

Millennial participants consistently emphasized practical and efficiency motivations for using AI tools. A central theme in their reflection was the perception that "every day new tools emerge that can help simplify or improve the processes we already do". This reflects a generational orientation toward continuous technological exploration and openness to adopt innovations that enhance productivity.

One of the strongest motivations expressed was time-saving. Participants noted that AI allows them to "save time searching for certain information" and "complete tasks more quickly". Beyond speed, they also highlighted the ability of AI to generate ideas, assist in planning and support decision-making, thereby expanding its role from a passive tool to an active collaborator in work and learning contexts.

Millennials also viewed AI as an educational resource, mentioning its usefulness in "learning new things" and "obtaining information" for both professional and personal purposes. Importantly, they consider AI not only as a productivity enhancer but also as a life simplifier, aligning with their desire to streamline complex tasks and reduce everyday friction.

Finally, the motivation extended to economic efficiency, as participants identified the potential for AI to deliver savings in both time and money. Collectively, these insights suggest that millennials are motivated by a combination of pragmatism, curiosity and a results-oriented mindset, seeing AI as a tool that helps them manage multiple demands while maximizing resources.

2.4.3.4. *Millennials perceived risks of AI tools*

Despite the positive aspects of AI tools that were clearly stated by Generation Y participants, there are also some concerns that were evident in a participant's narratives. An important concern among users regarding the use of AI tools is the potential loss of cognitive and social abilities. Many participants expressed concerns about how people, especially younger users, can rely on AI technologies and this can lead to a decrease in critical thinking skills, diminished interpersonal interactions and a decrease of human creativity.

By thinking in the future, some participants expressed the fear of becoming dependent on AI for problem-solving tasks, and this raises concerns about the long-term impact on individuals' cognitive and social development. Generation Y users expressed this fear in terms of younger generations more than their own generation, as they reflected that students now will rely on AI tools for their academic assignments and therefore, it will prevent them from actually learning and thinking for themselves.

Because Generation Y users are at a stage of their life in which they are focused on work, the concern about how AI may impact their industries and businesses in the future is latent. Specifically, the perceived risk of job displacement due to AI automation ranks high among users' concerns. Particularly, there were certain professions that may become obsolete, leading to the replacement of humans in some jobs. For these reasons, they stressed the importance of viewing AI tools as a complement rather than a replacement for jobs and tasks, and that companies still need the "human touch" to interpret AI information and decide the best ways to use it.

With respect to the control users have with AI tools, participants expressed apprehension regarding the lack of control and transparency in AI tools' decision-making processes. Many participants worry about AI systems making biased or unethical decisions, leading to unintended consequences and harm to individuals or communities. The lack of knowledge in AI algorithms and the absence of regulations for their use, can increase fears of losing control over the outcomes generated by AI tools.

Despite most participants agreeing on the fact that AI tools can provide good information and answers to most of their questions, there are still concerns about the reliability and accuracy of information provided by AI tools – in all cases and contexts, but participants question the credibility of AI-generated information, in contexts where misinformation, manipulation or bias can have significant implications.

In general, in some cases, participants perceive AI as a double-edged sword, capable of driving innovation and disruption while also raising concerns about trust, and future implications. The paradox of trust and distrust underscores the need for more knowledge and regulations to build confidence in AI technologies, and promote responsible usage. As AI continues to permeate various aspects of daily life, efforts to enhance transparency, accountability and ethical standards will be crucial in shaping consumer perceptions and fostering widespread acceptance and adoption.

2.4.4. *Centennials*

2.4.4.1. *Centennials' understanding of AI*

In the case of understanding AI, the centennials gave a very general description of its use. The responses varied between those who considered that AI has become a recurring topic, that they have recently seen it in movies and that it has had a practical application in academic projects in which they have participated. They also highlighted that the use of AI refers to the way in which technology learns autonomously

> Artificial intelligence, well, it has already become very, as they would say, mainstream... you find it everywhere and they talk about it everywhere (Rodrigo, centennial).

> I've heard that it's the way, basically, that machines learn (Fernanda, centennial).

Most of the examples provided by the young people who were interviewed focused on the popular ChatGPT platform. It is also important to note that the use of this application had an academic connotation in practically all cases. This can be explained because the interviewees were all university students:

> When I don't understand a problem, I can use GPT chat to verify it and it explains the procedure to me if my teacher hasn't explained it to me (Fernanda, centennial).

> ChatGPT is a resource to have information, not to copy works. And that is the one I know the most right now, there are other projects from other computers that work like this, but it is the one I know the most (Lucia, centennial).

2.4.4.2. *Perceived usefulness*

Some of the advantages of using AI for centennials are quick information, efficiency, as well as the ease of doing work, as noted before. The most common uses are academic works in most cases, with applications such as ChatGPT. This last platform received mentions in most cases:

> Reduce the time you spend on one activity, support yourself with it and be able to do another, because that is already a lot of time, a lot of efficiency. If they are there, it is very clear, that is, it is well programmed there and whatever, that is a lot of efficiency (Manuel, centennial).

> Generating content with artificial intelligence reduces the labor, so to speak, of creating articles and putting them or videos or multimedia content to users and does not generate as much. Well, time to create them, to publish them and so on (Enrique, centennial).

Despite the perception that the use of AI has a generally positive effect on the new generations, some were skeptical about its use. There was concern that the quality of the work being done by their peers would end up being of low quality:"

> I think it is something that can be used a lot in our favor if you know how to use it, it can also make the average human become mediocre with this new artificial intelligence (Mariana, centennial).

> It is a tool that can be used in good and bad ways, for easy things like giving your ideas or laborious things that make you lazy, but I think

> that at a certain level it is an abuse when they use it to do assignments, theses, final projects (Maximiliano, centennial).

2.4.4.3. *Perceived risks*

In the case of perceived risks, the centennial users consulted mentioned that dependence on this technology could have undesirable consequences. They mentioned that it could lead to the theft of information, as well as that it could have catastrophic consequences with a reserved prognosis. But they also expressed concern that users would become more vulnerable to marketing strategies that use this technology:

> Artificial intelligence makes a much faster analysis of many variables and many behaviors that not even the people themselves have the use of reason that have this type of behavior and that it is carried out by their subconscious, which is why it makes us more consumerist (Paola, centennial).

> Information is a fundamental risk that being able to share your data, that you are aware of what data you are transmitting (Valeria, centennial).

2.5. Discussion and conclusion

This chapter compared cross-generational perspectives on AI among baby boomers, Generation X, millennials and centennials (Gen Z). Using a qualitative design based on in-depth interviews, the study explored the understanding, perceptions of usefulness, motivations to use and perceived risks of AI in these four generations. The findings revealed clear generational contrasts in how AI is interpreted and used.

In line with prior studies on technology adoption, such as [KIM 21] and [SUD 23], our findings show that across generations, AI is largely perceived as useful and functional, but the benefit is contingent on how individuals integrate these technologies into their routines.

This aligns with the uses and gratifications theory, where the value of technology depends not only on its availability but also on the user's perceived needs, expectations and gratifications they obtain by its use. With this context, the following discussion aims to synthesize how different generations perceive the overall AI use (see Table 2.1).

	Baby Boomers	Gen X	Millennials	Gen Z
Understanding of AI	-Prevalent and complex topic. -Science fiction becoming a reality. -Robots that can replace human intelligence. -Machine, system or algorithm that searches for answers based on information on the web.	-Machines and systems that mimic human thinking, learning and communication. -Science fiction becoming real. -Many equate any form of automation with AI.	-Mimicks human learning and decision-making. -Tool for handling complex and routine tasks. -Driver of innovation and disruption. -Enhances efficiency and saves time. -Mixed feelings: trust in usefulness, but concern over reliability and transparency.	-View AI as a normal, integrated part of daily life. -Engage with AI intuitively due to lifelong exposure to digital technologies. -Understand AI through personal use (e.g. social media algorithms, recommendation systems).
Motivations to use	-Facilitation of daily tasks and activities. -Assistance in professional activities. -Solve complex world problems. -Learn new things.	-Facilitation of daily tasks and activities. -Learning and information seeking. -Curiosity and fear of missing out (FOMO).	-Facilitate complex and routine tasks. -Saving time and costs.	-Curiosity and eagerness to explore AI's creative and interactive potential. -Personalization, instant results and convenience. -Expression, learning and digital identity formation.
Perceived usefulness	-Optimization of tasks and increased efficiency. -Easy and inexpensive access to information. -Potential to solve complex global problems.	-Boosts performance. -Saves time, cost, effort. -Aids decision-making, reduces errors. -Improves access to precise info. -Useful in daily and work tasks. -Supports learning and skill development. -Enables global communication, tackles social/environmental issues.	-Facilitating processes and activities. -Optimizing, efficiency, time saving, and cost saving. -Convenience and accuracy.	-Essential for communication, education, entertainment, and decision-making. -Value AI for time management, content curation, and productivity. -Use AI tools to simplify tasks, discover new interests, and stay connected.

	Baby Boomers	Gen X	Millennials	Gen Z
Perceived risks	-Lack of adequate controls or regulations. -Diminish human intelligence. -Loss of socialization abilities and community. -Too old to adapt.	-Loss of cognitive and social skills. -Health risks from reduced mobility and interaction. -Difficulty discerning true versus false information. -Job displacement. -Loss of control and potential AI misuse. -Algorithmic bias and manipulation. -Privacy violations and identity theft. -Risks of espionage, extortion and human rights breaches.	-Loss of cognitive and social abilities. -Loss of jobs. -Lack of control of AI tools. -Quality of provided information.	-Privacy and surveillance. -Manipulation through algorithms and echo chambers. -Worry about overdependence and reduced critical thinking. -Question transparency and ethical use of their personal data.

Source: Developed by authors.

Table 2.1. *Comparison of AI understanding, usefulness, motivation and risk across generations*

2.5.1. *Baby boomers: cautious and curious*

Baby boomers recognized AI's potential in specific, targeted contexts. They consider AI can facilitate some tasks and offer assistance in professional activities. Further, they recognize the ease of use and inexpensive access to information. However, they maintained a cautious optimism, often highlighting concerns about complexity, ethical risks and the challenge of keeping up with fast-paced innovation. This reflects a generational stance that balances curiosity, skepticism and risk, grounded in the perceived effort required to adapt to new technological systems.

2.5.2. *Generation X: pragmatic realism*

Generation X exhibited a pragmatic and balanced perspective. While acknowledging the practical benefits of AI, in their words such as efficiency, personalization and convenience, they consistently raised concerns regarding privacy, data misuse and security. Additionally, they see the risk of humans being replaced by AI. Their view suggests a negotiation between embracing functionality and managing potential risks, positioning them as critical adopters of this technology.

2.5.3. *Millennials: enthusiastic yet concerned*

Millennials, as digital natives, displayed strong enthusiasm for AI's potential to enhance productivity, convenience and lifestyle integration. Their discourse reflects an expectation that AI can support both professional and personal spheres. At the same time, they voiced significant concerns over job displacement and data security, opening debate to broader societal debates about automation and labor, as well as the ethical handling of personal information. This duality highlights a generation deeply embedded in technology, but at the same time, aware of its disruptive potential.

2.5.4. *Centennials: acceptance with awareness*

Gen Z users, the youngest group, showed the highest acceptance and curiosity toward AI, often viewing it as an integral and normalized element of daily life. Their openness reflects early exposure and integration of smart technologies into their routines. Nonetheless, this group also expressed awareness of potential misuse, particularly anticipating negative outcomes related to personal safety and ethical boundaries [ROH 23, SOH 24]. Furthermore, they are also concerned about developing overdependence and, in consequence, reducing their critical thinking.

This suggests that while centennials are confident adopters, they are not uncritical, demonstrating an informed and reflexive approach to AI.

In summary, the findings indicate that while usefulness and functionality are mostly cross-generational constants, the types of risks and gratifications perceived vary by generation. These distinctions highlight the importance of tailoring AI communication, design and policy interventions to different generations.

In terms of managerial and theoretical implications, these findings underscore the importance of personalized approaches in promoting AI adoption and addressing each generation's concerns. By examining generational differences, organizations and policymakers can tailor strategies to address specific concerns, build trust and encourage the widespread adoption of AI technologies. Developers, educators and policymakers can leverage this generational diversity to design inclusive AI solutions aligned with varying user needs and priorities. This knowledge can also inform improvements in user-experience design, marketing strategies and educational initiatives aimed at reducing perceived risks across these generational cohorts. Ultimately, understanding generational perspectives on AI not only sheds light on the dynamics of its adoption but also highlights the importance of bridging generational gaps to foster a shared vision for ethical and effective AI integration into society.

Finally, as no research is without limitations, the present study considered a sample size of 39 participants, which provided rich qualitative insights, but may not be representative of the broader population. Furthermore, the data collection method is limited to interviews, which, though effective for in-depth exploration for a new topic such as this one, may benefit from incorporating other data collection methods, such as focus groups. Additionally, as perceptions of AI are likely to evolve rapidly with advancements in technology, the findings represent a snapshot in time and may not fully capture future generational shifts in attitudes. Future research should aim to address these limitations by expanding the sample size, incorporating diverse methodologies and exploring additional contextual factors.

2.6. References

[ADA 92] ADAMS D.A., NELSON R.R., TODD P.A., "Perceived usefulness, ease of use, and usage of information technology: A replication", *MIS Quarterly*, pp. 227–247, 1992.

[ADA 21] ADADI A., "A survey on data-efficient algorithms in big data era", *Journal of Big Data*, vol. 8, no. 24, pp. 1–54, 2021.

[BIT 20] BITKINA O.V., JEONG H., LEE B.C. et al., "Perceived trust in artificial intelligence technologies: A preliminary study", *Human Factors and Ergonomics in Manufacturing & Service Industries*, vol. 30, no. 4, pp. 282–290, 2020.

[CAL 19] CALVO-PORRAL C., PESQUEIRA-SANCHEZ R., "Generational differences in technology behaviour: Comparing millennials and Generation X", *Kybernetes*, vol. 49, no. 11, pp. 2755–2772, 2019.

[CAM 25] CAMPBELL ACADEMIC TECHNOLOGY SERVICES, AI in higher education: A meta summary of recent surveys of students and faculty, Report, Buies Creek, 2025.

[CHE 07] CHEN C.H., MORT G.S., "Consumers' technology adoption behaviour: An alternative model", *The Marketing Review*, vol. 7, no. 4, pp. 355–368, 2007.

[CHO 21] CHOI T.R., DRUMWRIGHT M.E., "OK, Google, why do I use you? Motivations, post-consumption evaluations, and perceptions of voice AI assistants", *Telematics and Informatics*, vol. 62, p. 101628, 2021.

[CRA 25] CRANE L., GREEN M., SOTO P., Measuring AI uptake in the workplace, FEDS Notes, Board of Governors of the Federal Reserve System, Washington, D.C., 2025.

[DAV 89] DAVIS F.D., "Perceived usefulness, perceived ease of use, and user acceptance of information technology", *MIS Quarterly*, pp. 319–340, 1989.

[DEL 23] DE LA GARZA D., *Interacciones en redes sociales y sus efectos en el mundo real: reflexiones sobre una década de profundos cambios sociales y tecnológicos*, Tirant lo Blanch, 2023.

[DHO 01] DHOLAKIA U.M., "A motivational process model of product involvement and consumer risk perception", *European Journal of Marketing*, vol. 35, no. 11/12, pp. 1340–1360, 2001.

[FIL 12] FILIPOVIĆ J., "Theory of uses and gratifications: A review", *Marketing Science*, vol. 43, pp. 219–228, 2012.

[HAS 21] HASAN R., SHAMS R., RAHMAN M., "Consumer trust and perceived risk for voice-controlled artificial intelligence: The case of Siri", *Journal of Business Research*, vol. 131, pp. 591–597, 2021.

[HIR 06] HIRUNYAWIPADA T., PASWAN A.K., "Consumer innovativeness and perceived risk: Implications for high technology product adoption", *Journal of Consumer Marketing*, vol. 23, no. 4, pp. 182–198, 2006.

[JEN 02] JENSEN K.B., "Media effects: Quantitative traditions", in *A Handbook of Media and Communication Research: Qualitative and Quantitative Methodologies*, Routledge, London, pp. 138–155, 2002.

[JON 22] JONES V.K., "Why people use virtual assistants: Understanding engagement with Alexa", *Journal of Brand Strategy*, vol. 11, no. 1, pp. 80–101, 2022.

[KAT 74] KATZ E., BLUMLER J., GUREVITCH M., "Utilization of mass communication by the individual", in BLUMLER J. and KATZ E. (eds), *The Uses of Mass Communications*, Sage, Beverly Hills, CA, 1974.

[KEL 23] KELLY S., KAYE S.A., OVIEDO-TRESPALACIOS O., "What factors contribute to the acceptance of artificial intelligence? A systematic review", *Telematics and Informatics*, vol. 77, p. 101925, 2023.

[KIM 21] KIM J., MERRILL K., COLLINS C., "AI as a friend or assistant: The mediating role of perceived usefulness in social AI vs. functional AI", *Telematics and Informatics*, vol. 64, p. 101694, 2021.

[MOR 20] MORTON F., "Influencer marketing: An exploratory study on the motivations of young adults to follow social media influencers", *Journal of Digital & Social Media Marketing*, vol. 8, no. 2, pp. 156–165, 2020.

[NAS 15] NASIR M.A., WU J., YAGO M. et al., "Influence of psychographics and risk perception on internet banking adoption: Current state of affairs in Britain", *International Journal of Economics and Financial Issues*, vol. 5, no. 2, pp. 461–468, 2015.

[PAL 16] PALUCH S., WÜNDERLICH N.V., "Contrasting risk perceptions of technology-based service innovations in inter-organizational settings", *Journal of Business Research*, vol. 69, no. 7, pp. 2424–2431, 2016.

[RAA 08] RAACKE J., BONDS-RAACKE J., "MySpace and Facebook: Applying the uses and gratifications theory to exploring friend-networking sites", *CyberPsychology & Behavior*, vol. 11, no. 2, pp. 169–174, 2008.

[ROH 23] ROHDEN S.F., ZEFERINO D.G., "Recommendation agents: An analysis of consumers' risk perceptions toward artificial intelligence", *Electronic Commerce Research*, vol. 23, no. 4, pp. 2035–2050, 2023.

[SAD 22] SADRIWALA M.F., SADRIWALA K.F., "Perceived usefulness and ease of use of artificial intelligence on marketing innovation", *International Journal of Innovation in the Digital Economy*, vol. 13, no. 1, p. 1–10, 2022.

[SHI 11] SHIN D.H., "Understanding e-book users: Uses and gratification expectancy model", *New Media & Society*, vol. 13, no. 2, pp. 260–278, 2011.

[SOH 24] SOHN S., "Consumer perceived risk of using autonomous retail technology", *Journal of Business Research*, vol. 171, p. 114389, 2024.

[STA 04] STAFFORD T.F., STAFFORD M.R., SCHKADE L.L., "Determining uses and gratifications for the Internet", *Decision Sciences*, vol. 35, no. 2, pp. 259–288, 2004.

[STA 25] STANFORD HUMAN-CENTERED ARTIFICIAL INTELLIGENCE, Artificial Intelligence Index Report, 2025.

[SUD 23] SUDARYANTO M.R., HENDRAWAN M.A., ANDRIAN T., "The effect of technology readiness, digital competence, perceived usefulness, and ease of use on accounting students artificial intelligence technology adoption", *E3S Web of Conferences*, vol. 388, p. 04055, 2023.

[THA 21] THANGAVEL P., PATHAK P., CHANDRA B., "Millennials and Generation Z: A generational cohort analysis of Indian consumers", *Benchmarking: An International Journal*, vol. 28, no. 7, pp. 2157–2177, 2021.

[TRE 15] TREVIÑO T., The online world of digital moms: Motivations, meanings, and role of brands in their everyday lives, Doctoral Thesis, EGADE Business School, Instituto Tecnológico y de Estudios Superiores de Monterrey, 2015.

[TRE 19] TREVIÑO T., PINEDA GARELLI J.L., "Understanding digital moms: Motivations to interact with brands on social networking sites", *Qualitative Market Research: An International Journal*, vol. 22, no. 1, pp. 70–87, 2019.

[TUN 17] TUNDREA D., Artificial intelligence is getting personal: A study on the usage motivations and privacy concerns of intelligent personal assistants' users, Thesis, Uppsala University, 2017.

[VEN 17] VENTER E., "Bridging the communication gap between Generation Y and the Baby Boomer generation", *International Journal of Adolescence and Youth*, vol. 22, pp. 497–507, 2017.

[WAL 23] WALD R., PIOTROWSKI J.T., ARAUJO T. et al., "Virtual assistants in the family home: Understanding parents' motivations to use virtual assistants with their children", *Computers in Human Behavior*, vol. 139, p. 107526, 2023.

[WHI 13] WHITING A., WILLIAMS D., "Why people use social media: A uses and gratifications approach", *Qualitative Market Research: An International Journal*, vol. 16, no. 4, pp. 362–369, 2013.

3

Smart Decisions: How AI Is Transforming Everyday Management and Engineering Practices

Artificial intelligence (AI) is no longer a futuristic concept; it is rapidly becoming an everyday ally in decision-making across management and engineering fields. This chapter demystifies AI by breaking down complex ideas into practical insights tailored for professionals without a technical background. It explores how AI-driven tools are enhancing productivity, precision and creativity in domains ranging from business analytics and human resources to project management and engineering design. With an emphasis on no-code and low-code platforms, readers are introduced to intuitive, web-based AI applications that require minimal technical know-how but deliver substantial impact. Ethical considerations and organizational readiness are also addressed, equipping readers with the confidence to navigate AI adoption responsibly. The chapter concludes with a forward-looking perspective, encouraging a culture of curiosity and continuous learning as the key to making smarter, empowered decisions in a digitally evolving world.

3.1. Introduction

The nature of decision-making is undergoing a fundamental transformation (Scott 2024). In an age defined by complexity, speed and digital interconnectivity, professionals are increasingly required to make decisions with unprecedented levels of data, uncertainty and consequence (Samala and Rawas 2025). Traditional methods, reliant on human intuition, experience and static models, are no longer sufficient to navigate today's dynamic environments. Whether managing operations

Chapter written by Soha RAWAS, Cerine TAFRAN, Agariadne Dwinggo SAMALA, Feri FERDIAN and Yudha Aditya FIANDRA.

in a global supply chain or designing resilient infrastructure in uncertain conditions, professionals are turning to artificial intelligence (AI) as a means to augment human judgment with data-driven precision, predictive foresight and operational agility (Qiu et al. 2025).

AI, once perceived as a distant frontier of computer science, has now entered the mainstream of professional practice (Faulconbridge et al. 2025). It is embedded in tools we use every day, often unnoticed, shaping how we plan, communicate, analyze and innovate (Bialkova 2024). From smart recommendation engines and automated scheduling assistants to AI-enhanced design software and intelligent dashboards, the influence of AI is subtle yet profound. It supports decision-making across sectors, accelerates routine tasks and opens new possibilities for creativity and strategy (Madhumita et al. 2024). Crucially, it does so without requiring every user to understand machine learning (ML) algorithms or write a single line of code.

For managers, AI enables real-time insights into customer behavior, workforce performance and market shifts, empowering leaders to act faster and with greater confidence (Vatamanu and Tofan 2025). For engineers, AI enhances planning accuracy, improves simulation fidelity and optimizes resource allocation (Abbas et al. 2025). Yet despite these benefits, a significant gap persists between what AI can offer and how widely it is understood or adopted, particularly among professionals who do not come from technical or data science backgrounds. Much of this reluctance stems from a perception that AI is inherently complex, opaque or reserved for a narrow group of specialists (Rawas 2024a).

This chapter challenges that notion. It presents AI not as a disruptive force that replaces human intelligence, but as a practical, approachable tool that extends it. It argues that AI is no longer an optional add-on for forward-thinking organizations; it is becoming a core competency for professionals who seek to remain relevant, resilient and innovative. And importantly, it emphasizes that the path to adopting AI does not require technical mastery, but rather a shift in mindset: from resistance to curiosity, from fear to experimentation.

To support this shift, this chapter offers a clear and grounded understanding of what AI is, and what it is not, paired with real-world illustrations and intuitive tools that demonstrate its practical value in everyday management and engineering contexts. It also addresses the ethical and organizational implications of AI adoption, underscoring the importance of responsible use, transparency and human-centered design. By the end of the chapter, readers will be equipped not only with a foundational awareness of AI's capabilities but also with the confidence to explore its applications, adopt its tools and lead change in their respective domains.

3.2. What is AI? A practical overview

AI is often portrayed as a mysterious, futuristic force, associated with humanoid robots, sentient machines or hyper-advanced algorithms operating beyond human comprehension (Jaboob et al. 2024). These portrayals, though popular in media, have contributed to widespread misconceptions that obscure AI's true nature and practical value. In reality, AI is not magic, nor is it confined to laboratories or large tech firms. It is, at its core, a set of computational techniques that enable machines to mimic certain aspects of human intelligence, particularly learning, reasoning, pattern recognition and decision-making (Krishna 2024; Batool et al. 2025).

At a practical level, AI refers to systems that can process large volumes of data, identify patterns and make informed predictions or decisions without being explicitly programmed for every scenario (Dobrin 2025). This includes machine learning (ML), a subset of AI in which algorithms improve automatically through experience (Bell 2022). For example, a ML model can analyze historical sales data to forecast future demand or detect anomalies in sensor readings that signal potential equipment failure (Ahaggach et al. 2024).

AI is also embedded in systems we interact with daily, often in subtle ways (Hirvonen et al. 2024; Prasad and Makesh 2024). When Netflix recommends a movie based on your past viewing history, or when Google Maps reroutes your trip to avoid traffic, AI is at work. In these examples, AI is not a robot with consciousness; it is a data-driven mechanism designed to enhance convenience, efficiency and personalization.

Crucially, modern AI tools are becoming increasingly user-friendly, abstracting away technical complexity through no-code or low-code interfaces (Chapelle 2024; Rawas 2024b; Selvakumar et al. 2024). This means that managers, engineers, educators and professionals in various fields can now leverage AI without a deep background in computer science. Tools such as ChatGPT can draft emails, summarize documents or even generate creative content in seconds. Platforms such as Power BI or Google Looker Studio integrate AI-powered features to visualize data trends and derive actionable insights with minimal configuration.

Despite these advances, several misconceptions still persist (Giray 2024; El Fathi et al. 2025; Gaitantzi and Kazanidis 2025). One of the most common is that AI equals automation, or that AI systems are inherently autonomous and decision-making entities in their own right. In reality, most AI applications today are narrow AI; they are trained for specific tasks (e.g. recognizing images, translating languages, segmenting customer data) and operate under human-defined goals and boundaries. Another myth is that AI requires coding expertise to use; while that may be true for developers building the underlying models, most professionals can now access AI's capabilities through simple interfaces designed for practical use.

Understanding AI as a spectrum, ranging from simple rule-based automation to sophisticated learning systems, helps to situate it more realistically within professional settings. It is not an all-or-nothing technology; rather, it is a continuum of tools that can be applied incrementally, depending on the problem, the context and the user's needs.

By reframing AI as a supportive technology rather than a complex or threatening one, this section sets the stage for exploring how AI is being applied across management and engineering contexts, not as a replacement for professional expertise, but as an augmentation of it. The next sections demonstrate exactly how this augmentation is taking shape in everyday workflows through real-world use cases and accessible platforms.

3.3. AI for smarter management practices

In the era of digital acceleration, managers are expected to make faster, more informed decisions across increasingly complex operational landscapes. The traditional reliance on intuition and retrospective analysis is being replaced, or more precisely, augmented, by data-driven intelligence (Valle-Cruz and García-Contreras 2025). AI is no longer a luxury reserved for tech giants; it is now embedded into the everyday decision-making fabric of modern organizations (Al-Raeei 2025). Through intuitive, no-code platforms, managers can access insights, streamline workflows and automate tasks that previously demanded significant time, manpower and technical expertise (Agrawal 2023).

3.3.1. *The expanding role of AI in management*

AI is transforming core managerial functions, ranging from performance analysis and talent acquisition to customer engagement and strategic planning, through increasingly accessible, no-code tools. As shown in Figure 3.1, these technologies are now embedded across all major functional areas of management. In business analytics, AI enables predictive dashboards, trend forecasting and interactive reports. Human resources benefit from AI-driven resume screening, interview analysis and automated shortlisting. Customer engagement is enhanced through ChatGPT-based chatbots, personalized replies and support automation. In marketing, AI supports campaign optimization, content generation and audience segmentation. Project management is streamlined with AI-powered scheduling, automated task routing and dynamic risk prediction. Finally, executive planning is empowered by scenario simulation and AI-assisted goal-setting. This comprehensive integration not only boosts speed and accuracy but also democratizes advanced insights, allowing non-technical managers to lead more effectively and make smarter decisions with confidence.

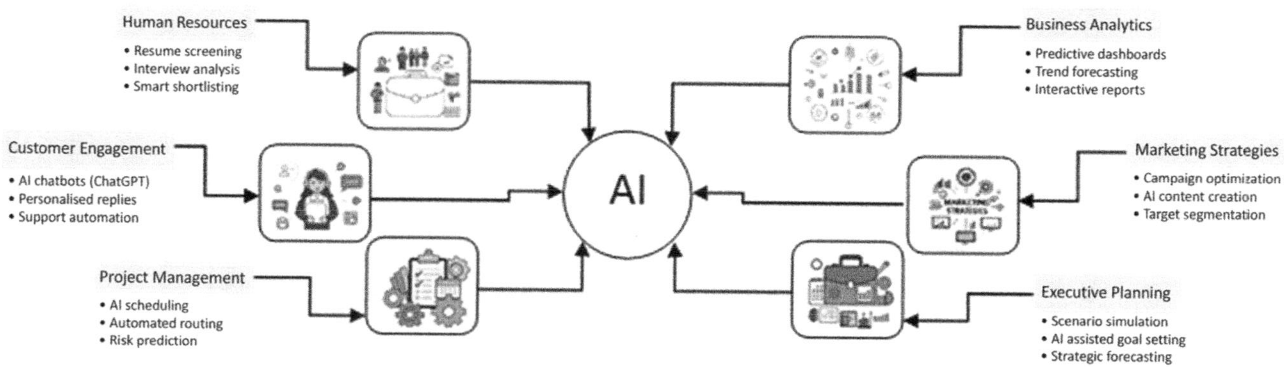

Figure 3.1. *Functional areas of management enhanced by AI. For a color version of this figure, see www.iste.co.uk/machado/Artificial.zip*

This ecosystem of intelligent tools not only improves speed and accuracy but also democratizes insights, allowing non-technical managers to lead with confidence.

3.3.2. *Real-world tools and use cases*

Table 3.1 presents a curated comparison of widely adopted AI tools that are reshaping management practices, emphasizing the ease of use and accessibility (Islam 2023).

Domain	Tool	Use case	AI feature	Ease of use
Business analytics	Google Looker Studio	Real-time KPI dashboards, sales reports	Smart visualizations, predictive analysis	Drag-and-drop
	Tableau Public	Interactive charts, trend analysis	Natural language queries, clustering	Moderate
HR and recruitment	Recruitee	Resume parsing, applicant tracking	AI-based filtering and scoring	Guided workflows
	HireVue	Video interview evaluation	Emotion and tone analysis (ethically debated)	Medium
Marketing	Canva Magic Write	Generating marketing copy	Prompt-based content creation	Very easy
	Jasper AI	Long-form marketing campaigns	Template-based AI writing	Easy
Customer support	ChatGPT	Drafting responses, chatbot backend	Generative dialogue modeling	Text-based prompts
Project planning	Notion AI	Meeting summaries, timeline generation	Text summarization, decision support	Seamless integration

Table 3.1. *Accessible AI tools for everyday management functions*

These platforms are designed to minimize the learning curve, with most offering web-based access, intuitive UIs and integrations with widely used enterprise software. For example, a customer service manager can use ChatGPT to draft responses in seconds or create an automated script for common queries, while a HR officer might screen dozens of applications within minutes using Recruitee's AI filters, without writing a single line of code.

3.3.3. *Strategic advantages of AI-augmented management*

The integration of AI into managerial workflows offers significant strategic advantages:

– Time efficiency: automates data collection, reporting and repetitive communications.

– Decision quality: enhances accuracy with predictive and real-time insights.

– Scalability: allows lean teams to perform high-volume tasks effectively.

– Consistency: standardizes outputs such as email responses, evaluations and recommendations.

– Innovation: frees cognitive space for managers to focus on vision, culture and long-term planning.

Critically, these advantages are attainable without deep technical knowledge. AI acts as a silent partner, supporting the manager's role, not replacing it.

3.4. AI in engineering: enhancing efficiency without coding

The engineering landscape is undergoing a paradigm shift as AI becomes embedded in design, operations and planning processes (Schwalbe and Finzel 2024). Once confined to highly technical or research-heavy environments, AI is now available to engineers through intuitive, visual platforms that require little to no programming (Hopgood 2021). This democratization of AI means that engineers can focus on innovation and efficiency, using AI to simulate scenarios, optimize resource use, predict failures and automate tedious tasks, without needing to become data scientists (Zong and Guan 2024).

3.4.1. *How AI is reshaping engineering workflows*

AI's role in engineering spans multiple domains, mechanical, civil, electrical and industrial, through a range of intelligent features embedded in everyday tools (Alenizi et al. 2023). From design automation to predictive maintenance, these applications help reduce trial-and-error, improve accuracy and support proactive decision-making.

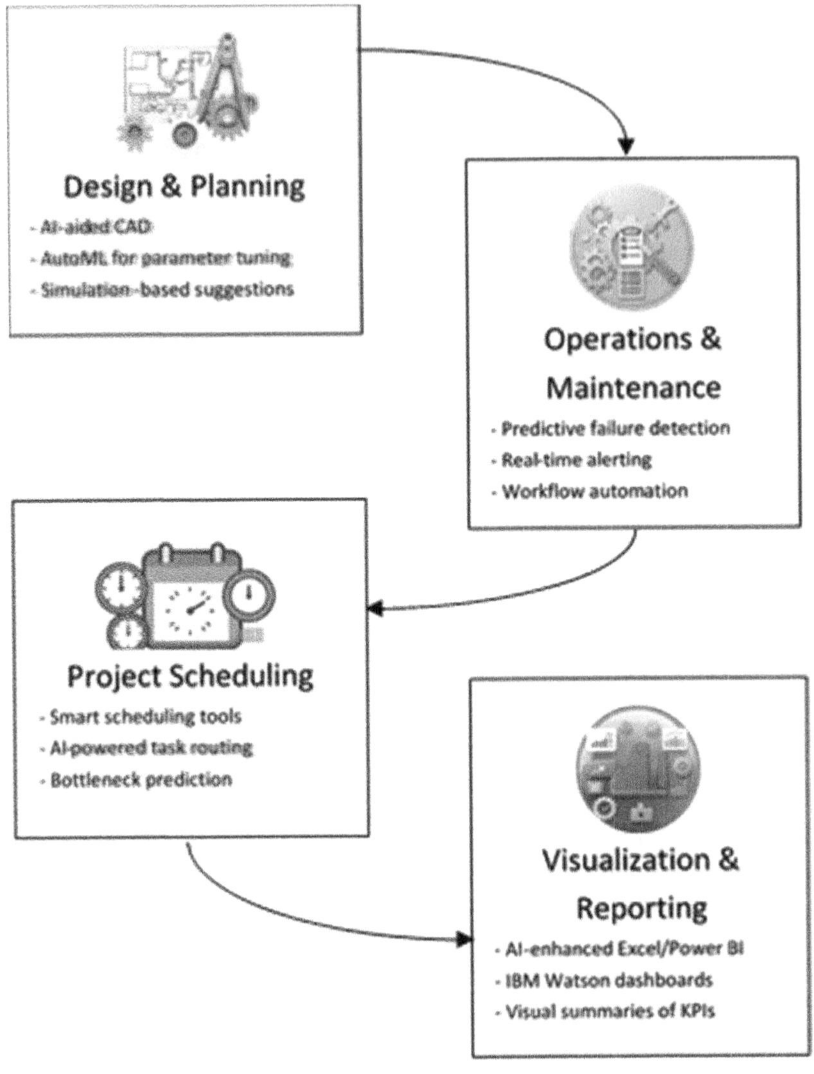

Figure 3.2. *AI applications across the engineering lifecycle. For a color version of this figure, see www.iste.co.uk/machado/Artificial.zip*

Figure 3.2 shows how AI is reshaping engineering workflows across the entire lifecycle. In the Design & Planning phase, tools such as AutoML support parameter tuning and simulation-based optimization, while AI-aided CAD provides design suggestions. During Operations & Maintenance, AI enables predictive failure detection, real-time alerts and workflow automation. In Project Scheduling, smart tools assist with task prioritization, bottleneck prediction and dynamic resource allocation. Lastly, the Visualization & Reporting stage benefits from AI features in platforms such as Excel Copilot and IBM Watson, offering predictive dashboards, KPI summaries and enhanced reporting. Together, these applications demonstrate how engineers can use intuitive, no-code AI platforms to drive innovation and efficiency without requiring programming expertise.

This lifecycle view shows how AI contributes to faster development cycles, optimized asset usage and improved forecasting, even when used by professionals with no formal AI training.

3.4.2. *No-code AI tools for engineering use cases*

Several platforms now offer AI features tailored to the needs of engineering professionals, without requiring users to write code or build models from scratch (Alenizi et al. 2023), as shown in Table 3.2.

Application area	Tool	Functionality	AI feature	Usability level
Design and simulation	Google AutoML	Auto-generates models for optimization and predictions	Model training without coding	Easy to Moderate
	Autodesk Generative Design	Suggests optimal design alternatives	AI-powered CAD logic	Built-in CAD UI
Project scheduling	Monday.com with an AI assistant	Task prioritization, resource allocation	Smart recommendations	Drag-and-drop + AI
	Asana + AI	Suggests deadlines, risk alerts, capacity balance	Timeline optimization	Intuitive Dashboard
Reporting and dashboards	Microsoft Excel + Copilot	Data summaries, trend analysis, predictive suggestions	Natural language insights	Familiar interface
	IBM Watson Studio	Predictive analytics, dashboard building	Integrated ML for engineers	Moderate (guided)

Table 3.2. *Practical AI tools for engineering efficiency*

These tools empower engineers to harness AI's analytical strength without needing to understand the underlying algorithms. For example, using Autodesk Generative Design, a mechanical engineer can input desired constraints, such as strength, material use and cost, and receive optimized structural designs generated by the AI engine. Similarly, Excel's AI-based Copilot can instantly summarize complex datasets and suggest chart types for engineering reports.

3.4.3. *Key advantages in engineering contexts*

AI adoption in engineering delivers strategic and operational advantages:

– Design optimization: AI algorithms can generate thousands of design variations, meeting performance and cost constraints with minimal manual input.

– Proactive maintenance: predictive models help forecast failures or bottlenecks, enabling timely intervention and minimizing downtime.

– Smarter scheduling: AI helps balance resource loads, assign tasks and adapt plans dynamically in large-scale projects.

– Data simplification: AI automates data interpretation, allowing engineers to focus on outcomes rather than raw metrics.

Most importantly, these benefits are increasingly accessible to individual engineers and small teams, not just to large enterprises or AI specialists. By integrating AI seamlessly into existing tools and platforms, professionals can amplify their capabilities without technical overload.

3.5. Easy-to-use AI tools for non-technical professionals

The democratization of AI has reached a point where powerful AI capabilities are now accessible without specialized training or technical skills (Borger et al. 2023). For managers, engineers and professionals across disciplines, this means AI is no longer confined to data scientists or programmers; it is a practical resource that can be leveraged daily to enhance productivity, creativity and decision-making.

Modern AI tools focus on usability, minimal setup and integration with familiar workflows (Ullah et al. 2025). Most operate via web browsers, require no installation and offer intuitive interfaces that enable users to perform complex tasks through simple prompts or clicks. This shift has made AI accessible for drafting emails, summarizing meetings, generating multimedia content, and more, activities that previously required significant time and effort (Gama and Magistretti 2025).

3.5.1. *Key user-friendly AI platforms*

Table 3.3 summarizes some of the most popular AI tools designed specifically for non-technical professionals, highlighting their primary functions and user advantages.

Tool	Primary use case	Key features	User experience
ChatGPT	Drafting emails, reports and content	Conversational AI, content generation, Q&A	Text prompts; instant responses
Notion AI	Meeting summaries, note-taking, planning	AI-powered summaries, task suggestions	Embedded in familiar workspace UI
Lumen5	Automated video creation	Converts text to video with visuals and music	Drag-and-drop interface
Pictory	Video summarization for presentations	Extracts key points and creates short videos	Upload text or video; AI editing
Tome	Smart presentations	AI-generated slide decks with narrative flow	Web-based; intuitive storytelling

Table 3.3. *Accessible AI tools for everyday professional use*

3.5.2. *Practical applications across roles*

– Managers can use ChatGPT to draft clear, professional emails or reports, saving hours of writing and editing time. Notion AI can summarize lengthy meeting notes and suggest action items, improving team communication.

– Engineers and technical professionals may leverage Lumen5 and Pictory to create engaging videos explaining complex designs or project updates without the need for video editing skills.

– Marketing and communications teams benefit from Tome's AI-driven presentations, enabling rapid production of compelling slide decks tailored to diverse audiences.

3.5.3. *Advantages of no-code AI tools*

– Accessibility: most tools require only basic digital literacy and no prior AI knowledge.

– Speed: tasks that took hours, such as content drafting, video creation or summarization, can be completed in minutes.

– Cost-effectiveness: many platforms operate on freemium models or affordable subscriptions, lowering barriers for individuals and small teams.

– Integration: these tools often work seamlessly with popular productivity suites (e.g. Google Workspace, Microsoft 365), enhancing existing workflows.

By empowering professionals to use AI easily and effectively, these tools foster a culture of experimentation and innovation. Users can test AI capabilities in low-risk environments, gaining confidence to expand AI adoption within their organizations.

3.6. Ethical and organizational considerations

As AI becomes an integral part of professional workflows, embracing its benefits must be accompanied by a clear commitment to ethical principles and organizational readiness. Responsible AI adoption is not merely a technical challenge; it involves careful attention to transparency, fairness, privacy and the human impact of automated decisions. Managers and engineers, often the primary users or implementers of AI tools, play a crucial role in ensuring that AI serves organizational goals without unintended harm or bias.

3.6.1. *Transparency and explainability*

One of the foundational ethical concerns with AI is transparency, the ability to understand how AI systems arrive at their conclusions or recommendations (Balasubramaniam et al. 2023). While many AI tools operate as “black boxes”, organizations must strive for explainability, especially when AI influences decisions affecting people, such as hiring or customer service. Users should demand clear documentation and insight into AI decision-making processes and be prepared to question and verify outputs rather than accepting them at face value.

3.6.2. *Bias and fairness*

AI systems learn from historical data, which can embed societal biases, leading to unfair or discriminatory outcomes (Pulivarthy and Whig 2025). For example, AI-based resume screening tools may inadvertently favor certain demographic groups if trained on unbalanced datasets. It is vital for organizations to audit AI tools regularly, assess their impact on diverse groups and implement corrective measures to mitigate bias. Ethical AI use means prioritizing fairness, inclusivity and respect for all stakeholders.

3.6.3. *Data privacy and security*

AI's power depends on data, often large volumes of sensitive information. Protecting this data is paramount. Organizations must comply with data protection regulations (such as GDPR or CCPA), ensure secure handling and storage of data, and maintain user confidentiality (Saeed et al. n.d.). Transparent data policies help build trust among employees, customers and partners, fostering a culture of responsible AI use.

3.6.4. *Organizational readiness and culture*

Successful AI adoption requires more than technology; it demands organizational readiness. This involves educating teams about AI capabilities and limitations, promoting open communication about AI-driven changes and encouraging a culture of experimentation balanced with caution (Greene and Burton 2025). Leaders should foster psychological safety where employees feel empowered to raise concerns or suggest improvements related to AI tools.

Change management strategies that include clear guidelines, training and ethical frameworks are critical to reduce fear and resistance. Preparing the workforce to collaborate effectively with AI systems enhances both acceptance and outcomes.

3.6.5. *Evaluating ethical AI tools*

Before integrating AI tools, managers and engineers should evaluate them for ethical design principles:

– Does the tool provide transparency about its algorithms and data sources?

– Are there mechanisms to detect and correct bias?

– How does the tool protect user data and privacy?

– Is there support for human oversight and intervention?

Choosing AI solutions aligned with these criteria helps ensure that technology supports organizational values and social responsibility.

3.7. Future outlook: embracing AI with confidence

AI is no longer a nascent technology confined to specialized labs or elite organizations; it is rapidly becoming an integral component of professional life across industries and disciplines. As AI tools grow more sophisticated, accessible and embedded within everyday workflows, the question is no longer *if* professionals should engage with AI, but *how* they can do so effectively and responsibly.

Looking ahead, several key trends will shape the landscape of AI in management and engineering:

– Continued democratization: AI capabilities will become even more embedded within familiar platforms, reducing the need for specialized skills and making AI tools accessible to a broader range of professionals. This will enable individuals and teams to innovate without technical barriers.

– Hybrid human–AI collaboration: rather than replacing human expertise, AI will increasingly augment it. Professionals who combine domain knowledge with AI insights will lead the way in making smarter, faster and more nuanced decisions.

– Ethical and regulatory maturation: as AI adoption grows, so will frameworks for ethical use, data governance and accountability. Staying informed about evolving best practices and regulations will be essential for sustainable AI integration.

– Lifelong learning and adaptability: the rapid pace of AI innovation requires a mindset of continuous learning. Professionals who cultivate curiosity, stay updated on AI developments and experiment with new tools will maintain a competitive edge.

To thrive in this evolving environment, organizations and individuals alike must foster a culture that encourages curiosity, experimentation and openness. Encouraging small-scale pilots, cross-disciplinary collaboration and ongoing education can reduce fear and build confidence around AI adoption.

Moreover, staying connected with professional communities, attending workshops and leveraging trusted AI literacy resources can empower professionals to distinguish meaningful innovation from hype. Importantly, embracing AI with confidence does not mean abandoning critical thinking or ethical vigilance; it means integrating AI thoughtfully as a complement to human insight.

In sum, the future of professional decision-making is not AI replacing humans, but humans empowered by AI, equipped to make smarter, more agile and more informed decisions in an increasingly complex world.

3.8. Conclusion

Artificial intelligence is no longer an abstract concept or distant future technology; it is a practical, accessible and powerful tool reshaping the everyday work of managers and engineers alike. Throughout this chapter, we have seen that AI is not about replacing human judgment but about enhancing it, amplifying our ability to analyze data, automate routine tasks and make smarter decisions with confidence and speed.

By demystifying AI concepts and highlighting no-code and low-code tools, this chapter has aimed to empower professionals across disciplines to embrace AI without fear or technical overwhelm. The examples and platforms discussed demonstrate that AI's value lies in its usability and adaptability, enabling users to integrate intelligent capabilities into their workflows regardless of prior technical expertise.

Equally important are the ethical and organizational considerations that accompany AI adoption. Responsible use, transparency, fairness and data privacy must be foundational to any AI strategy, ensuring that technology serves people and aligns with organizational values.

Looking ahead, the key to success lies in cultivating a culture of curiosity and continuous learning, where experimentation with AI tools is encouraged, and professionals are prepared to evolve alongside emerging technologies. Smart decisions, after all, come from empowered people who combine domain knowledge with AI-driven insights, leveraging the best of both human and machine intelligence.

As AI continues to evolve, the opportunity is clear: by embracing accessible AI tools and thoughtful practices today, professionals can lead their organizations into a future defined by innovation, agility and smarter decision-making.

3.9. Declarations

During the preparation of this work, the authors used generative AI tools (e.g. ChatGPT) to assist in refining the language, structuring the document and improving clarity. The core ideas, scientific analysis, methodology design, implementation and all critical thinking were solely the work of the authors.

3.10. References

Abbass, M., Akhai, S., Abbas, U., Jafri, R., Arif, S.M. (2025). AI-enabled sustainable urban planning and management. In *Real-World Applications of AI Innovation*, Mallik, S., Mathivanan, S., Sangeetha, S., Soufiene, B. (eds). IGI Global Scientific Publishing, Hershey, PA, 233–260. doi: 10.4018/979-8-3693-4252-7.ch012.

Agrawal, A. (2023). *No-Code Artificial Intelligence: The New Way to Build AI Powered Applications*, English edition. BPB Publications, New Dehli.

Ahaggach, H., Abrouk, L., Lebon, E. (2024). Systematic mapping study of sales forecasting: Methods, trends, and future directions. *Forecasting*, 6(3), 502–532.

Al-Raeei, M. (2025). The smart future for sustainable development: Artificial intelligence solutions for sustainable urbanization. *Sustainable Development*, 33(1), 508–517.

Alenizi, F.A., Abbasi, S., Mohammed, A.H., Rahmani, A.M. (2023). The artificial intelligence technologies in Industry 4.0: A taxonomy, approaches, and future directions. *Computers & Industrial Engineering*, 185, 109662. doi: 10.1016/j.cie.2023.109662.

Balasubramaniam, N., Kauppinen, M., Rannisto, A., Hiekkanen, K., Kujala, S. (2023). Transparency and explainability of AI systems: From ethical guidelines to requirements. *Information and Software Technology*, 159, 107197. doi: 10.1016/j.infsof.2023.107197.

Batool, A., Zowghi, D., Bano, M. (2025). AI governance: A systematic literature review. *AI and Ethics*, 1–15.

Bell, J. (2022). What is machine learning? In *Machine Learning and the City: Applications in Architecture and Urban Design*, Carta, S. (ed.). Wiley, Hoboken, NJ, 207–216.

Bialkova, S. (ed.) (2024). AI transforming business and everyday life. In *The Rise of AI User Applications: Chatbots Integration Foundations and Trends*. Springer, Cham, 143–165.

Borger, J.G., Ng, A.P., Anderton, H., Ashdown, G.W., Auld, M., Blewitt, M.E., Brown, D.V., Call, M.J., Collins, P., Freytag, S. et al. (2023). Artificial intelligence takes center stage: Exploring the capabilities and implications of ChatGPT and other AI-assisted technologies in scientific research and education. *Immunology & Cell Biology*, 101(10), 923–935. doi: 10.1111/imcb.12689.

Chapelle, C.A. (2024). Open generative AI changes a lot, but not everything. *Modern Language Journal*, 108(2). Wiley, Hoboken, NJ.

Dobrin, S.I. (2025). *Talking about Generative AI: A Guide for Educators 2.0*. Broadview Press, Peterborough.

El Fathi, T., Saad, A., Larhzil, H., Lamri, D., Al Ibrahmi, E.M. (2025). Integrating generative AI into STEM education: Enhancing conceptual understanding, addressing misconceptions, and assessing student acceptance. *Disciplinary and Interdisciplinary Science Education Research*, 7(6). doi: 10.1186/s43031-025-00125-z.

Faulconbridge, J., Sarwar, A., Spring, M. (2025). How professionals adapt to artificial intelligence: The role of intertwined boundary work. *Journal of Management Studies*, 62(5), 1991–2024.

Gaitantzi, A. and Kazanidis, I. (2025). The role of artificial intelligence in computer science education: A systematic review with a focus on database instruction. *Applied Sciences*, 15(7), 3960.

Gama, F. and Magistretti, S. (2025). Artificial intelligence in innovation management: A review of innovation capabilities and a taxonomy of AI applications. *Journal of Product Innovation Management*, 42(1), 76–111.

Giray, L. (2024). Ten myths about Artificial Intelligence in education. *Higher Learning Research Communications*, 14(2), 1–12.

Greene, B.A. and Burton, S.L. (2025). Organizational readiness for artificial intelligence (AI) in network security. *Organizational Readiness and Research: Security, Management, and Decision Making*, 205–246.

Hirvonen, N., Jylhä, V., Lao, Y., Larsson, S. (2023). Artificial intelligence in the information ecosystem: Affordances for everyday information seeking. *Journal of the Association for Information Science and Technology*, 75(10), 1152–1165. doi: 10.1002/asi.24860.

Hopgood, A.A. (2021). *Intelligent Systems for Engineers and Scientists: A Practical Guide to Artificial Intelligence*. CRC Press, Boca Raton, FL.

Islam, S.M.R. (2023). *AI in Everyday Life: Tools and Applications*. Amazon KDP (Independently published).

Jaboob, A., Durrah, O., Chakir, A. (2024). Artificial intelligence: An overview. In *Engineering Applications of Artificial Intelligence*, Chakir, A., Andry, J.F., Ullah, A. et al. (eds). Springer, Cham, 3–22.

Krishna, V.V. (2024). AI and contemporary challenges: The good, bad and the scary. *Journal of Open Innovation: Technology, Market, and Complexity*, 10(1), 100178.

Madhumita, G., Das, T., Das, S., Khatri, E., Ravisankar, P., Hemachandu, P. (2024). IoT and AI for real-time customer behavior analysis in digital banking. In *2024 5th International Conference on Recent Trends in Computer Science and Technology (ICRTCST)*, Jamshedpur, 198–203. doi: 10.1109/ICRTCST61793.2024.10578458.

Prasad, R. and Makesh, D. (2024). Impact of AI on Media & Entertainment Industry. In *Media & Journalism Transformations-Emerging Trends and Paradigm Shifts*, Pandey, A., Pravin, T.R., Prasad, R.K. (eds). Emerald Publications, Bingley, 41–71.

Pulivarthy, P. and Whig, P. (2025). Bias and fairness addressing discrimination in AI systems. In *Ethical Dimensions of AI Development*, Bhattacharya, P., Hassan, A., Liu, H., Bhushan, B. (eds). IGI Global, Hershey, PA, 103–126.

Qiu, K., Puccinelli, N., Ciniselli, M., Di Grazia, L. (2025). From today's code to tomorrow's symphony: The AI transformation of developer's routine by 2030. *ACM Transactions on Software Engineering and Methodology*, 34(5), 1–17. doi: 10.1145/3709353.

Rawas, S. (2024a). AI: The future of humanity. *Discover Artificial Intelligence*, 4(1), 25.

Rawas, S. (2024b). ChatGPT: Empowering lifelong learning in the digital age of higher education. *Education and Information Technologies*, 29(6), 6895–6908.

Saeed, M.M., Saeed, R.A., Ali, E.S., Ahmed, Z.E., Mokhtar, R.A. (2025). Privacy and security challenges of AI integration in Industry 4.0 and 5.0. In *Computational Intelligence for Analysis of Trends in Industry 4.0 and 5.0*, Awotunde, J.B., Muduli, K., Brahma, B. (eds). Auerbach Publications, New York, 353–372. doi: 10.1201/9781003533023-14.

Samala, A.D. and Rawas, S. (2025). Bias in artificial intelligence: Smart solutions for detection, mitigation, and ethical strategies in real-world applications. *International Journal of Artificial Intelligence*, 14, 32–43.

Schwalbe, G. and Finzel, B. (2024). A comprehensive taxonomy for explainable artificial intelligence: A systematic survey of surveys on methods and concepts. *Data Mining and Knowledge Discovery*, 38(5), 3043–3101.

Scott, W. (2024). *AI Everyday: Transforming Lives with Smart Technology*. eBookit.com.

Selvakumar, P., Muthusamy, S., Satishkumar, D., Vigneshkumar, P., Selvamurugan, C., Satheesh Kumar, P. (2024). AI-powered tools. In *Using Real-Time Data and AI for Thrust Manufacturing*, Satishkumar, D. and Sivaraja, M. (eds). IGI Global, Hershey, PA, 20–42.

Ullah, F., Saqib, S., Xiong, Y.-C. (2025). Integrating artificial intelligence in biodiversity conservation: Bridging classical and modern approaches. *Biodiversity and Conservation*, 34(1), 45–65.

Valle-Cruz, D. and García-Contreras, R. (2025). Towards AI-driven transformation and smart data management: Emerging technological change in the public sector value chain. *Public Policy and Administration*, 40(2), 254–275.

Vatamanu, A.F. and Tofan, M. (2025). Integrating Artificial Intelligence into public administration: Challenges and vulnerabilities. *Administrative Sciences*, 15(4), 149.

Zong, Z. and Guan, Y. (2025). AI-driven intelligent data analytics and predictive analysis in Industry 4.0: Transforming knowledge, innovation, and efficiency. *Journal of the Knowledge Economy*, 16, 864–903. doi: 10.1007/s13132-024-02001-z.

4

Integrating AI into Business Education: Bridging the Gap Between Disciplinary Knowledge and Business Performance

Artificial intelligence (AI) has revolutionized the way businesses operate, enabling efficient automation of routine tasks and freeing up human resources for more strategic and creative activities. In the educational sphere, business schools are adopting AI to enhance the academic experience and develop essential skills in students. However, incorporating AI into business schools poses challenges and opportunities that require a systematic approach to consider AI literacy, critical thinking, creativity, ethics and how AI need to be understood from a disciplinary angle. This exploratory study aims to address key research questions, including how business schools respond to industry needs and ensure competency development when curricula is incorporating AI. A conceptual model is proposed that discusses the proper utilization of AI in terms of necessary skills and incremental application of tools according to disciplinary competencies. Additionally, the importance of evaluating student learning outcomes and perceptions about the relevance of AI education to their future careers is highlighted.

4.1. Introduction

Artificial intelligence (AI) has emerged as a transformative force in business by integrating advanced algorithms, machine learning (ML) and data analytics. Organizations have unlocked unprecedented potential for optimizing processes, making more intelligent decisions and creating innovative products and services. In essence, AI provides a paradigmatic shift in the way businesses operate, enabling efficient automation of routine tasks and freeing up human resources for more

---

Chapter written by Laura Esther Zapata CANTÚ and Martha Elena Moreno BARBOSA.

strategic and creative activities. This not only increases operational efficiency but also fosters an environment conducive to innovation, allowing teams to focus on solving complex challenges and exploring new opportunities. In this regard, firms must consider additional investments and innovations that will be critical to provide workers with the skills and training required to work with AI systems and to qualify for the new tasks and occupations they create over time.

Additionally, the integration of AI in business processes demands a new profile of leaders who are not only tech-savvy but also possess the ability to foster innovation, collaboration in AI-enhanced environments and "soft skills" such as emotional intelligence, critical thinking and adaptability. This holistic approach to skill development ensures that employees are not only technically proficient but also equipped to handle the complex interpersonal and ethical challenges posed by AI technologies.

Technological advances in business have also impacted the way institutions of higher education develop or adapt their academic program curricula. Historically, curriculum design was shaped by what institutions sought to promote or market within their discipline, typically emphasizing foundational topics and current trends. From the 1990s to the present, orientation to competence development has grown exponentially. The curriculum progressively being altered by industry needs regarding abilities and performance. During this period competences became central, whereas previously technology had been used primarily as a tool for content delivery. Today, the introduction of AI is not limited to providing content, it encompasses understanding content and reinforcing the development of disciplinary competence.

The introduction of AI into business schools transcends mere technological advancement and extends beyond traditional teaching methods, reshaping the educational experience. AI enhances educational processes and develops essential skills, challenging business schools when considering offering AI courses due to the lack of pedagogical resources, as well as to ensure the development of disciplinary competencies.

The typical teaching methods in management and business courses rely on case studies, students cannot fully grasp the impact of AI in organizations by merely reading stories about how companies and organizations use AI to aid in decision-making and problem-solving. It is unlikely business graduates will need to handle all the intricacies of AI algorithms and representations. AI training in business schools may not necessarily require programming assignments, as is the case in computing and engineering disciplines (Xu and Babaian 2021). In business schools, the integration of AI in the classroom necessitates a systematic approach that considers AI literacy, critical thinking, creativity and ethics.

In this regard, examining the implications, challenges and opportunities of incorporating AI courses in shaping the future of business education became critical. This integration requires a paradigm shift in how education is approached using AI, moving beyond traditional methods to embrace more dynamic, interactive and student-centered learning environments (Chiu et al. 2022).

Assessing student learning outcomes and perceptions is of equal importance, as it can serve as an essential input for curriculum design and improvement. However, since AI courses have not been widely offered in business schools, there has been little research reporting student learning outcomes and their opinions towards various aspects of the curriculum, as well as their perceptions of the relevance of AI education to their future careers (Xu and Babaian 2021).

In this sense, we believe a gap is growing between the requirements of organizations and the competencies that universities are developing. Through this exploratory study, we intend to respond to two main research questions: (Almasri 2024) How do business schools respond to industry needs by incorporating AI? (2) How do business schools ensure the development of competencies, including soft skills and disciplines, when AI is incorporated?

Several scholars are calling for new studies to broaden the understanding of AI's impact on students and academics (Shum and Luckin 2019). For instance, Kumar et al. (2024)) argue that "the exclusive research on the application of AI in business education is relatively novel compared to other fields and requires the attention of researchers and academicians" (Kumar et al. 2024).

4.2. AI in business practices and education

Considering the irruption of AI in business organizations, nowadays it has also become relevant to understand AI benefits and challenges to contrast with the development of disciplinary competencies. In this regard, the literature presents how AI has disrupted organizations' processes, enhancing the way collaborators work and workplaces must adapt accordingly. At the same time, AI is revolutionizing higher education by offering tools and technologies that enhance the learning process and experiences. This section presents how emerging technologies, such as AI, are transforming organizations and business schools. It is currently imperative to work together to face the challenges that technological advances bring.

4.2.1. *AI boosting the use of technology in workplaces*

AI is a collection of information and communication technologies (ICTs) that imitate human intelligence for the primary purpose of enhancing jobs, increasing efficiencies and driving economic growth (Arakpogun et al. 2021). The intelligent agents (IAs) that result from AI activities hold a large amount of know-how that is required to improve productivity and create new knowledge for business processes.

An AI-driven approach, for instance, is a strategy whereby IA enables the accessibility of valuable information via technology-driven platforms for employees. Furthermore, IA has a wide range of capacities in contributing to the organization's approaches for innovation through strategic knowledge activities. In this regard, the implementation of AI in business requires a comprehensive approach that considers the enablers and inhibitors that can influence its adoption and effective use. AI enablers and inhibitors can be categorized into three levels: technological, organizational and environmental (Enholm et al. 2022).

Technological infrastructure is a key factor for AI implementation, as it requires a trained team, suitable algorithms and the ability to handle large amounts of data. In this sense, technological readiness is crucial to ensure the successful adoption of AI. The organizational context also plays a crucial role in the adoption of AI. Organizational structure, strategy and culture can either drive or hinder the successful implementation of AI. Executives must lead the implementation of AI and ensure it is not limited to the IT department. Additionally, employees must understand the purpose of AI, their role in its implementation and how it will impact their responsibilities within the organization.

Harari (2024), in his study on the evolution of technology uses in today's society, shows us that the explosion of AI was a predictable feature, given the advance in capacity and mobility made possible by networks (speed) and the ease of access to data of different types (variety) and in different places (volume), which would be adding value to various organizational and social actions, as long as the veracity of the data is ensured (Harari 2024).

In this context, the big data cycle emerges (Davenport and Kirby 2016), offering immense learning possibilities to neural networks and intelligent algorithms (Rezaei 2025). In this regard, it is estimated that the use of AI has made excellent contributions to addressing sensitive areas, with the potential to assist people who have visual, motor and even emotional difficulties (Vaswani et al. 2017).

For example, people with difficulties can use applications and artifacts that are assistive devices with AI software, which, when combined with glasses, make content of various types accessible, providing relief for people with disabilities and those with visual impairments (Enholm et al. 2022).

AI applications must consider both ethical and moral principles, as well as regulatory and policy frameworks, during the development and implementation of AI in organizations. By considering these factors, companies can ensure successful AI implementation and leverage its benefits to improve efficiency and competitiveness in the market. (Enholm et al. 2022)

4.2.2. *AI in business practices*

In the era of AI, machine learning models are trained on large datasets. The ability of AI to process large volumes of information in real-time enables more accurate predictive analysis, the identification of patterns and trends, and valuable insights to guide strategic decisions. For organizations that are aware of data analysis and handle tasks such as data discovery, big data, and data preparation and integration, AI is a natural next step. A mature analytics system will be the foundation of AI success. AI systems mature over time as they are fed more quality data. Therefore, businesses invest in data storage and data warehousing; this is part of the process of aligning assets to implement AI effectively.

Many functions within a business can benefit from analytics. The most common functional categories include:

– Customer analytics: this category includes applications to marketing and customer experience.

– Supply chain analytics: this includes demand forecasting and optimization of inventory, pricing, scheduling, transportation and storage, while mitigating risk.

– Human capital analytics: this subfield applies to service industries where human resources are the primary means of production.

– Fraud and risk analytics: this includes assessment of several types of risk (market, operational, credit), particularly in the financial sector.

While many aspects of a business can be automated and should be, tasks that require judgment, prioritization and trade-offs still necessitate human intelligence. The integration of AI into specific business processes where it can produce positive results is a good starting point. Over time, it will become natural to integrate AI into any new business initiative.

For instance, an area where AI artifacts are particularly effective is interior design, especially when combined with virtual reality (Wilsing et al. 2024). This author speculates that there are visible advantages to incorporating ML, especially for image processing, overcoming outdated color catalogs and devising architectural varieties such as building designs adaptable to changing climate conditions (Rajendra 2019).

To determine the best strategy, businesses should consider whether the task at hand allows for repetition, high volume, a predictable pattern and a low error cost. Tasks that satisfy these specific criteria are starting points for AI implementation. AI can be used to provide more effective responses to transaction requests by efficiently collecting and analyzing historical chat data.

By understanding the potential of AI and its applications in business, companies can harness its power to drive innovation and growth. Effective AI implementation requires a strategic approach that considers the organization's goals, data assets and talent. By following these guidelines, businesses can unlock the full potential of AI and achieve success in an increasingly competitive market (Rajendra 2019).

4.2.3. *AI as a driving force for business schools*

Nevertheless, challenges, limitations, permissiveness and excesses persist in the incorporation of this technology into learning and the way work is done in these segments. Such concerns involve skills training, competency development, teaching and pedagogical models, assessments and feedback, which, according to Santaella (2023), are still in their embryonic stages.

Therefore, some commentators emphasize the potential end of critical thinking (Lo 2023), reflected, for example, in the initial ban in 2020, on its use at the University of Michigan. Others adopt a more flexible stance, such as King (2023) who envisions the use of the technology within an environment of ubiquitous content production. Thus, according to the author, it is necessary to rethink traditional pedagogical practices and incorporate ChatGPT® (and other AI systems) into everyday school and academic life.

AI systems, such as chatbots, can increase students' cognitive activity and reduce the gap between high- and low-performing students (Lee et al. 2023), empowering students to predict their learning outcomes and regulate their behavior strategically (Almasri 2024), identify gaps in learning and improve academic performance by linking with innovative assessment practices (Leite 2025). AI can positively influence academic performance and student satisfaction (Ouyang and Jiao 2021).

The integration of AI in business education is a growing area, with research exploring its benefits and challenges (Xu and Babaian 2021). Recent research has focused on the adoption and use of AI as an educational and pedagogical technology to enhance students' learning experience, such as personalized tutoring systems, educational robots and adaptive learning environments.

The learning experience research has also identified AI paradigms in education: "AI-direct, learner-as-recipient", where AI leads learning with a defined pathway for the student; "AI-supported, learner-as-collaborator", where AI optimizes interactions between students, information and technology; and "AI-empowered, learner-as-leader", where AI enhances learners' intelligence through a complex system (O'Connor 2023). However, there are authors who caution against the intrinsic limitations of AI, highlighting the importance of considering implications beyond AI's cognitive processes and calculations (Cope et al. 2021).

Despite the potential benefits and the growing enthusiasm, AI education in business schools faces more challenges than in engineering schools (Stine et al. 2019). Among these challenges are faculty development, course updates and pedagogical techniques. Faculty pressure is haunting the HE ecosystem mainly because content created by GenAI cannot be reliably detected by conventional plagiarism mechanisms, and not even the exact source can be indicated to show where the content was copied from (e.g. Farrelly and Baker 2023). Adding complexity to this discussion is the identified "behavior" of GPTs generating false and misleading text. Van Dis et al. (2023) found that an article generated by GenAI presented convincing responses, albeit with factual errors and incorrect data interpretation. This scenario raises awareness of the "thin line" that separates the benefits and harm of GenAI in the students' educational journey in higher education (Van Dis et al. 2023).

The difficulty in teaching introductory AI courses, even in engineering programs (Stine et al. 2019), stems from their complexity, which encompasses many advanced topics and techniques. In addition, AI research and applications are rapidly updating the body of knowledge related to AI, making the subject even more challenging to teach.

Finally, the scarcity of pedagogical resources and the lack of design principles and guidelines for curriculum development (Xu and Babaian 2021). Disciplinary associations, such as the Association to Advance Collegiate Schools of Business

(AACSB), have anticipated the potential impacts of AI on business education, but have not recommended a specific curricular model.

4.2.4. *AI in business education and curriculum alignment: disciplinary competencies*

In business schools, AI is recognized as a transformative force, and emerging trends in the industry are moving to the adaptation of academic programs, enabling students to graduate with relevant skills and be prepared to face the challenges of the labor market (Harry and Sayudin 2023). Integrating AI into curricula ensures that students acquire future-proof competencies and skills, enabling them to respond effectively to the demands of a constantly evolving business environment.

Furthermore, AI facilitates teaching innovation by enabling data-driven instructional design, simulation-based learning and interactive content delivery (Tominc and Rožman 2023). These innovations are crucial in disciplines such as management, marketing and finance, where digital transformation is reshaping industry norms.

In this regard, it becomes crucial to examine how AI technologies are being integrated into undergraduate business programs and what institutional strategies ensure that AI use aligns with pedagogical objectives, ethical standards and industry requirements. The literature presents several case studies conducted as the primary exploratory research strategy. The main results are oriented to understand the experiences of students and professors in using AI. For instance, in this study in Brazil, thirty-five engineering and business students used GenAI tools to enhance their group project activities. The results indicate that AI has enhanced collaboration among students and had a positive impact on their teamwork. Additionally, topics such as ethics, trust and the human role in the integration of GenAI have been addressed (Leite 2025).

Similar results are presented in the authors and references who conducted a study at a university in Romania. These authors investigated the knowledge and perception of students, professors and entrepreneurs regarding the use of AI in business education. The survey revealed that, although many students had previously used AI, there was a widespread lack of preparedness to adopt AI into the educational process. The results showed that only 71.3% of students used AI in business education, and 46.8% stated they were not prepared for its adoption. Furthermore, it was found that professors were less prepared than students to integrate AI into business education. The study highlighted the importance of the professors' role in the effective implementation of adaptive technologies in the educational process and emphasized the need to improve knowledge and understanding of AI tools and their benefits, thereby better preparing professors for their entrepreneurial roles.

Given the existing literature's suggestion of a lack of effective integration of AI as a teaching and learning methodology in business schools. The present study aims to understand the following questions: How do business schools respond to industry needs by incorporating AI? How do business schools ensure the development of competencies, including soft skills and disciplines when AI is incorporated?

4.3. Method

We use an exploratory research method to understand the dynamics between a curriculum design and industry needs, while, at the same time, AI has irrupted into industries and universities. As we have mentioned before, in industries, AI has shown several benefits; in universities, it has been used as a supportive teaching tool. A pedagogical methodology that incorporates specific AI tools aligned with industry benefits and disciplinary competences will be a booster to enhance and ensure not only the development of disciplinary competences but also digital and advanced technological competences.

This exploratory study was conducted in Mexican firms to consider the primary benefits and risks of using AI in their organizations. At the same time, it is relevant to understand the perception of these firms regarding business school curricula. From May to August 2025, an online questionnaire was sent to firms that have been working on several projects with a Mexican business school. The questionnaire was designed to explore the phenomenon, asking questions about the benefits and open-ended questions about the risks of AI in business practices, as well as students' competencies.

A total of 107 questionnaires were completed. The profile of the respondents is as follows: 60% of the respondents were over forty years old; 31% were between 30 and 39 years old, while only 9% were under thirty. The survey results primarily reflect the perspective of more experienced professionals

Regarding the sectors in which respondents work, 50% of the respondents are used by companies in the service sector. In comparison, there is also a significant representation from the manufacturing industry (37%) and, to a lesser extent, from trade (13%). Another relevant criterion was the country of origin for the firm: 57% of the respondents are from Mexico, followed by the United States at 19%, 13% are European, 9% are from Latin America, and only 2% are from Asia. This result could indicate a diverse organizational culture and a firm's openness to adopting emerging technologies.

Age Ranges	Absolute frequency	Relative frequency
Less than 30	**10**	**9%**
Less than 25	1	1%
25-29	9	8%
30-39	**33**	**31%**
30-34	13	12%
35-39	20	19%
40 or more	**64**	**60%**
40 - 44	25	23%
45 or more	39	36%
Grand Total	**107**	**100%**

Sector	Absolute frequency	Relative frequency
Manufacture	40	37%
Services	53	50%
Trading/Distribution Company	14	13%
Grand Total	**107**	**100%**

Origin	Absolute frequency	Relative frequency
Mexico	61	57%
United States	20	19%
Europe	14	13%
Latin America	10	9%
Asia	2	2%
Grand Total	**107**	**100%**

Company Size	Absolute frequency	Relative frequency
Small (Less than 49)	**22**	**21%**
Micro (0-9 employees)	13	12%
Small (10-49 employees)	9	8%
Medium (50-249 employees)	**13**	**12%**
Medium (50-249 employees)	13	12%
Large (250 or more)	**72**	**67%**
Large (250 or more)	72	67%
Grand Total	**107**	**100%**

Table 4.1. *Respondent and firm's profile*

Finally, Table 1 shows that many perceptions originate from corporate environments (67%), although the perspectives of small- and medium-sized organizations (33%) are also considered.

4.4. Results

4.4.1. *AI benefits in Mexican firms*

A descriptive analysis reveals that 17% of respondents consider the ability to quickly identify patterns, trends and opportunities through data analysis as the most relevant benefit, as well as the automation of repetitive tasks (17%).

Benefits	Absolute frequency	Relative frequency
Data analysis: AI can quickly analyze large amounts of data to identify patterns, trends, and opportunities.	**88**	**17%**
Automation of repetitive tasks: AI can automate tasks that require little or no creative thinking, freeing employees to focus on more strategic tasks.	**87**	**17%**
Process optimization: AI can identify areas for improvement in business processes, which can lead to increased efficiency and productivity	74	14%
Cost reduction: AI can help businesses identify and reduce costs in a variety of areas, such as automating tasks and optimizing processes	66	13%
Improved decision-making: By analyzing data, AI can help businesses make more informed and effective decisions.	58	11%
Better customer experience: AI can help businesses personalize the customer experience, which can improve customer satisfaction and loyalty.	42	8%
Innovation in products and services: AI can help companies develop new products and services, and improve existing ones.	41	8%
Accessing new markets: AI can help companies identify and access new markets.	31	6%
Increased cybersecurity: AI can help detect and prevent cybersecurity threats.	29	6%
Product and service innovation: AI can help companies develop new products and services, and improve existing ones.	5	1%
Grand Total	**521**	**100%**

Table 4.2. *Benefits of AI use for firms*

Another relevant insight is that the service and manufacturing sectors consistently highlight task automation and data analysis as the main benefits of AI. In contrast, marketing companies emphasize process optimization and cost reduction, demonstrating a more financial and operational focus

The results also emphasize the diversity of benefits that AI offers in Mexican organizations, highlighting its relevance and incorporation in key organizational activities and processes. In this regard, universities must understand the depth and breadth of the benefits of AI in business practices.

Benefits	Services	Manufacture	Trading / Distribution Company	Grand Total
Data analysis: AI can quickly analyze large amounts of data to identify patterns, trends, and opportunities.	**43**	**35**	10	88
Automation of repetitive tasks: AI can automate tasks that require little or no creative thinking, freeing employees to focus on more strategic tasks.	**44**	**35**	8	87
Process optimization: AI can identify areas for improvement in business processes, which can lead to increased efficiency and productivity	37	25	**12**	74
Cost reduction: AI can help businesses identify and reduce costs in a variety of areas, such as automating tasks and optimizing processes	29	26	**11**	66
Improved decision-making: By analyzing data, AI can help businesses make more informed and effective decisions.	27	23	8	58
Better customer experience: AI can help businesses personalize the customer experience, which can improve customer satisfaction and loyalty.	22	15	5	42
Innovation in products and services: AI can help companies develop new products and services, and improve existing ones.	19	16	6	41
Accessing new markets: AI can help companies identify and access new markets.	18	8	5	31
Increased cybersecurity: AI can help detect and prevent cybersecurity threats.	14	11	4	29
Product and service innovation: AI can help companies develop new products and services, and improve existing ones.	2	1	2	5
Grand Total	**255**	**195**	**71**	**521**

Table 4.3. *Benefits of AI use for firms by sector*

4.4.2. *AI risks in Mexican firms*

Lack of training in the use of AI tools is the most frequently reported risk among respondents, followed by the dehumanization of organizations.

Risks	Absolute frequency	Relative frequency
Labor – lack of training in the use of AI tools	**74**	**30%**
Cultural – dehumanization of organizations	64	26%
Economic – increase in efficiency and productivity leading to staff layoffs	43	17%
Social – excessive use of social apps	36	14%
Economic – strong investment in technology	32	13%
Grand total	**249**	**100**

Table 4.4. *Risks of AI use for firms by sector*

In services and manufacturing, the most frequently cited risks are a lack of training in AI use and the dehumanization of organizations. In the marketing sector (14 people), however, economic risks reach 57% (8/14), surpassing cultural risks, although a lack of training remains the main risk. In Mexico and the USA, a lack of training in AI predominates. In Europe, however, dehumanization is the most cited with 71% (10/14).

This result highlights the importance of organizations and universities incorporating, in addition to AI literacy, a holistic approach to strengthen and develop individual, interpersonal and ethical skills.

4.4.3. *Perception of competences requiring development in business schools*

Respondents mention that the main competencies students should develop are AI and data analytics, followed by critical thinking, creativity and innovation. Those respondents under thirty years old believe that developing critical thinking skills should be the primary focus. In comparison, respondents above thirty years old believe that technical skills in AI and data analytics are key.

Competencies	Absolute frequency	Relative frequency
Technical competence in AI and data analytics	**98**	46%
Critical thinking, creativity and innovation	65	31%
Ethical concern and regulation of AI	24	11%
Adaptability and continuous learning	24	11%
Grand total	**211**	**100%**

Table 4.5. *Firm requirements for graduate competency*

The respondent perceptions reveal a harsh reality: universities face significant challenges in devising an effective response to the introduction of ChatGPT and other AI applications, as well as their subsequent adoption by students, professors and academic collaborators. Whereas some institutions banned LLMs immediately, others embraced them wholeheartedly. As a result, universities must adopt the emergent technologies while simultaneously identifying the best approach through a specific methodology to assess the value of technology incorporation, its alignment with course objectives and student performance.

4.5. Discussion and conceptual model

The descriptive results obtained from the industry provide insights into the benefits that AI has offered to firms in Mexico and how business schools could take advantage of them to enhance their curriculum. Chen (2022) reveals there is a lack of studies on AI successfully developing and applying AI in the best and easiest path curriculum in business education (Sharma 2024). Employers are looking for skills, not specific courses of study, and degrees serve as a proxy for content knowledge (Stine et al. 2019). Students need to develop their critical thinking skills and problem-solving capabilities, with an understanding of how traditional industries are transforming their business models and cultures to incorporate extensive use of big data and analytics (Sharma 2024). Lack of training in the use of AI tools is the most frequently reported risk among respondents, followed by the dehumanization of organizations.

Companies can benefit from already established analytics capabilities and successfully develop and apply AI in the most effective and efficient path. Accordingly, AI as an analytics tool is naturally embedded in the data analytics curriculum in business schools (Sharma 2024). The AI curriculum in business schools is still in its early stages. The results also emphasize the diversity of benefits that AI has in Mexican organizations showing its relevance and incorporation in key organizational activities and processes. We can argue that business schools and universities as educational institutions must incorporate AI into their disciplinary contents.

Additionally, it is essential to strike a balance between technology and people in business practices and education. As the respondent mentions, the lack of training in the use of AI tools is the most frequently cited risk, as well as the dehumanization of organizations. Technology, with its disruptive forces, compels organizations to adapt and evolve. For instance, ML algorithms such as the one used for ChatGPT are inherently designed to learn from user data. This feedback loop, where users train the technology and the technology trains the users, can amplify the technology's intended or unintended consequences on people and society (Berthon et al. 2024; Sharma 2024).

Figure 4.1 shows how a comprehensive model is needed to integrate AI into the curriculum considering graduate competencies, the integration of AI in the curriculum, and what business organizations consider AI benefits. The objective is to foster the development of both soft and hard skills in students while also providing a deep understanding of the challenges and opportunities in the labor market. The gap between business organization and business schools could stretch if AI benefits and tools in specific business practices are not considered.

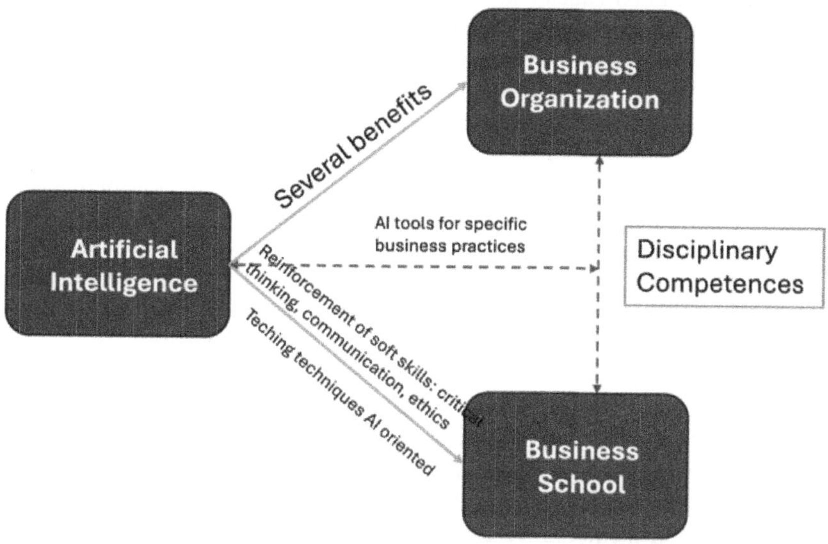

Figure 4.1. *Bridging the gap between business organizations and business schools*

Our results provide insights into addressing the research questions posed at the end of section 4.2.4. In the following sections, thoughtful insights provide the relevant elements to consider when incorporating AI for both business and universities.

4.5.1. *How do business schools respond to industry needs in the context of AI integration?*

The growing integration of AI in various professions has generated an imperative need to develop AI literacy curricula that not only foster technological proficiency but also encompasses competence in using AI in diverse contexts. In this sense, university students must have access to practical experience with industry-standard tools, which allows them to develop relevant and applicable skills in the workforce. However, the effective implementation of AI education faces significant challenges, including resource constraints and varying levels of access within educational institutions. To overcome these obstacles, strategic planning and partnerships are necessary to enhance accessibility and deliver a comprehensive AI learning experience.

Assessing problem-solving skills, algorithmic thinking and ethical considerations are key challenges in AI education. To address these challenges, a balanced approach is proposed that encompasses written exams, project evaluations and real-world applications. Continuous refinement of evaluation methods is crucial for adapting to the evolving nature of AI and providing meaningful feedback to students, thereby promoting a comprehensive understanding of AI concepts and applications.

The development of effective evaluation methods for AI education involves measuring both theoretical understanding and practical skills. Furthermore, establishing global standards for AI education is imperative to ensure consistency and quality across diverse educational institutions. Collaboration among international stakeholders is crucial for defining standard benchmarks and fostering a universally recognized framework for AI education.

The integration of AI in business schools offers significant opportunities to improve student training and prepare them for an increasingly automated future. However, it also poses significant challenges that require strategic planning and effective international collaboration. By addressing these challenges and developing effective evaluation methods, we can ensure that students receive high-quality AI education and are prepared to face the challenges of the 21st century (Sharma 2024).

4.5.2. *How do business schools ensure competence development (soft skills and disciplinary competences) when incorporating AI?*

AI has emerged as a transformative force in various fields, including business education. A recent study, conducted in 2019, with twenty-nine business schools selected based on their geographic location and current AI-related activities, highlights AI's ability to revolutionize businesses, industries and competitive environments, and, consequently, education in business schools.

This transformation requires a thorough review of curricula to integrate AI into academic programs, considering that new generations of business students have been exposed to AI-enabled education from primary school to university. However, the discrepancy in the life experience of business students can pose significant challenges.

The incorporation of AI in business education also raises questions about the role of faculty and the effectiveness of delivering learning in this model. The ability of business schools to respond to the rapidly evolving external environment largely depends on the leadership of the Dean and the university administration. A promising approach to addressing these challenges is the creation of an ecosystem that leverages the expertise of both faculty and industry professionals. By integrating industry experts who can provide up-to-date knowledge and applications, business schools can offer more relevant and practical education to students. This approach can also alleviate the pressure on faculty to be experts in all the latest technologies, allowing them to focus on their area of specialization.

Some business schools are establishing alliances to share their discoveries and best practices, reflecting a trend towards interdisciplinary and cross-campus collaboration to understand the multidimensional aspects of AI and its impact on industry, society and education. In this regard, business schools from various continents are forming alliances to develop new pedagogies and collaborate on academic programs.

A point of convergence among business schools is the importance of addressing the development of students' knowledge about AI from two complementary dimensions. On the one hand, there is a need to provide conceptual education on AI's capabilities and potential applications, allowing students to understand the fundamentals and practical uses of this technology. On the other hand, there is an emphasis on offering experiential education that includes both collaboration with AI systems (human–machine partnership) and the creation of AI systems, enabling students to apply their knowledge in practical contexts and develop relevant skills for the labor market.

Reviewing the literature exposes a clear understanding of crucial themes to be considered in the development of an educational innovative model, such as: a) harnessing AI for enhanced academic performance, b) AI ethics and trust impact on learning, c) AI as a supplement to human work, and d) ecosystem: main actors and their own actions and responsibilities.

In a new educational model, the addition of these themes requires a review and reinterpretation of course and classroom dynamics. Critical thinking and ethical considerations become essential. Critical thinking, in the context of AI education, involves the ability to analyze information, evaluate different perspectives and construct reasoned arguments within the framework of AI-driven environments. This skill is increasingly important as AI becomes more prevalent in various aspects of life and work. In educational settings, AI can be used as a tool not only for delivering content but also for encouraging students to question, analyze and think critically about the information they are presented with (Van den Berg and du Plessis 2023). The use of AI in education offers unique opportunities to cultivate critical thinking.

Even when the integration of AI into education offers significant benefits, it also raises important concerns. Scholars warn of risks such as data privacy violations, algorithmic biases and the displacement of human educators (Lampou 2023). Ensuring the ethical use of AI is critical. Classroom discussions are based on the recognition that AI is not omnipotent but has inherent limitations and weaknesses. Through direct experience of these limitations in practical tasks, students develop a greater appreciation for ethical considerations. Furthermore, they can analyze how the adoption of advanced AI technologies can generate significant and unprecedented risks to privacy, social equality, employment, public safety and global sustainability (Farrelly and Baker 2023).

Introducing ethics courses into academic training and building the capacity of AI development stakeholders can facilitate the integration of ethical values and the development of responsible AI (Kiemde and Kora 2022). The ethical implications of AI in education extend beyond technical considerations to encompass broader societal impacts, including privacy protection and social justice (Stine et al. 2019).

4.6. Conclusions

The contribution of our work is threefold: first, this exploratory study demonstrates the potential of providing business students with AI training, which is perceived as highly relevant to their future careers. It also presents a conceptual model that reinforces the incorporation of AI into the business school curriculum, emphasizing the development of disciplinary competencies. To achieve this, a specific methodology and pedagogical techniques must be designed to facilitate student learning. Second, we offer several practical guidelines for incorporating AI into the curriculum that other business educators may adopt in their courses. Third, it is essential to build or strengthen partnerships with industry to help universities and firms keep pace with the technological revolution.

4.6.1. *Theoretical implications*

AI's role in business skill enhancement is multifaceted, encompassing the development of both technical and soft skills. The studies reviewed here demonstrate the transformative potential of AI in reshaping workforce development, underscoring the need for innovative approaches to training and education in the digital era.

As businesses continue to integrate AI into their operations, the focus on skill enhancement will be crucial in maintaining a competitive edge and fostering a resilient, adaptable workforce. A new line of research is opening up to a better comprehension of how to bridge the gap between universities and organizations, incorporating emerging technologies.

Universities must consider two critical actions: on the one hand, to ensure that what AI promises is fulfilled in education delivery, including personalization, enhanced motivation and engaging students in their own learning process. On the other hand, training professors and collaborators to take advantage of its benefits. A complementary investment in human teaching assistants, human advisors and more universal access to reliable broadband and appropriate digital devices will be required to make such opportunities feasible for many adult learners.

4.6.2. *Practical implications*

Organizations operate in dynamic and constantly changing environments, composed of actors such as competitors and government, which influence how the organization can and should operate. AI applications must consider both ethical and moral principles, as well as regulatory and policy frameworks, during the development and implementation of AI in organizations. By considering these factors, companies can ensure successful AI implementation and leverage its benefits to improve efficiency and competitiveness in the market.

The results also highlight the diversity of benefits that AI offers in Mexican organizations, demonstrating its relevance and incorporation in key organizational activities and processes.

Business schools and universities, as a whole, educational institutions, must incorporate AI into their disciplinary content, and they also need to build alliances with industry. There is an invitation to bridge the gap by working together to face the challenges that technological advances bring – bridging the gap to develop new pedagogies and collaborate on academic programs that provide students with both theoretical foundations and hands-on experience. The alliance could include an understanding of AI's capabilities and applications, as well as engaging in collaborative projects with AI systems and contributing to their development. By balancing conceptual and experiential learning, business schools aim to equip students with the skills and knowledge needed for the evolving job market.

4.6.3. *Limitations*

Our study has several limitations. First, our exploratory study is based on Mexico, and the preliminary findings may not be generalizable to other country institutions. Second, we could not draw any statistical conclusions regarding the factors affecting learning outcomes due to the pilot stage of the study. As a result, our research is merely exploratory, and our findings may have limited applicability. Our future studies will seek to address the research questions in a large course sample.

Finally, AI usage in higher education is in its infancy but growing at a fast pace. Institutional support will be essential to ensure its sustainable and ethical use. Professional development should go beyond technical skills to include ethical considerations and strategies for enhancing learning. Additionally, collaborative communities will be vital for educators to share experiences and innovate collectively, enabling the effective and meaningful integration of GenAI.

4.7. Declaration

During the preparation of this work, the author(s) used generative AI Grammarly for grammar and spelling check.

4.8. References

Almasri, F. (2024). Exploring the impact of artificial intelligence in teaching and learning of science: A systematic review of empirical research. *Research in Science Education*, 54, 977–997. doi: 10.1007/s11165-024-10176-3.

Arakpogun, E.O., Elsahn, Z., Olan, F., Elsahn, F. (2021). Artificial Intelligence in Africa: Challenges and opportunities. In *The Fourth Industrial Revolution: Implementation of Artificial Intelligence for Growing Business Success*, Hamdan, A., Hassanien, A.E., Razzaque, A., Alareeni, B. (eds). Springer, Cham, 375–388.

Berthon, P., Yalcin, T., Pehlivan, E., Rabinovich, T. (2024). Trajectories of AI technologies: Insights for managers. *Business Horizons*, 67(5), 461–470.

Chan, C.K.Y. (2023). A comprehensive AI policy education framework for university teaching and learning. *International Journal of Educational Technology in Higher Education*, 38, 20. doi: 10.1186/s41239-023-00408-3.

Chen, L. (2022). Current and future artificial intelligence (AI) curriculum in business school: A text mining analysis. *Journal of Information Systems Education*, 33(4), 416–426.

Chiu, T.K.F., Xia, Q., Zhou, X., Chai, C.S., Cheng, M. (2022). Systematic literature review on opportunities, challenges, and future research recommendations of artificial intelligence in education. *Computers and Education: Artificial Intelligence*, 4, 202. doi: 10.1016/j.caeai.2022.100118.

Cope, B., Kalantzis, M., Searsmith, D. (2021). Artificial intelligence for education: Knowledge and its assessment in AI-enabled learning ecologies. *Educational Philosophy and Theory*, 53(12), 1229–1245.

Davenport, T.H. and Kirby, J. (2016). Just how smart are smart machines? *MIT Sloan Management Review*, 57(3), 21.

Davis, F.D. (1989). Perceived usefulness, perceived ease of use, and user acceptance of information technology. *MIS Quarterly*, 319–340. doi: 10.2307/249008.

Enholm, I.M., Papagiannidis, E., Mikalef, P., Krogstie, J. (2022). Artificial intelligence and business value: A literature review. *Information Systems Frontiers*, 24, 1709–1734. doi: 10.1007/s10796-021-10186-w.

Farrelly, T. and Baker, N. (2023). Generative artificial intelligence: Implications and considerations for higher education practice. *Education Sciences*, 13(11). doi: 10.3390/educsci13111109.

Goralski, M.A. and Tan, T.K. (2020). Artificial intelligence and sustainable development. *The International Journal of Management Education*, 18(1). doi: 10.1016/j.ijme.2019.100330.

Guha, A., Grewal, D., Atlas, S. (2023). Generative AI and marketing education: What the future holds. *Journal of Marketing Education*, 46(1), 6–17. doi: 10.1177/02734753231215436.

Harari, Y.N. (24 de August de 2024). Never summon a power you can't control: Yuval Noah Harari on how AI could threaten democracy and divide the world. *The Guardian* [Online]. Available at: https://www.theguardian.com/technology/article/2024/aug/24/yuval-noah-harari-ai-book-extract-nexus#:~:text=In%20recent%20generations%20humanity%20has, than%20ever%20to%20annihilating%20itself.

Harry, A. and Sayudin, S. (2023). Role of AI in education. *Injuruty: Interdisciplinary Journal and Humanity*, 2(3), 260–268. doi: 10.58631/injurity.v2i3.52.

Kiemde, S.M.A. and Kora, A.D. (2022). Towards an ethics of AI in Africa: Rule of education. *AI Ethics*, 2, 35–40. doi: 10.1007/s43681-021-00106-8.

Kumar, V., Ashraf, A.R., Nadeem, W. (2024). AI-powered marketing: What, where, and how? *International Journal of Information Management*, 77. doi: 10.1016/j.ijinfomgt.2024.102783.

Lampou, R. (2023). The integration of artificial intelligence in education: Opportunities and challenges. *Review of Artificial Intelligence in Education*, 4, e15–e15. doi: 0.37497/rev.artif.intell.educ.v4i00.15

Lee, M.C., Scheepers, H., Lui, A.K., Ngai, E.W. (2023). The implementation of artificial intelligence in organizations: A systematic literature review. *Information & Management*, 60(5). doi: 10.1016/j.im.2023.103816.

Leite, H. (2025). Artificial intelligence in higher education: Research notes from a longitudinal study. *Technological Forecasting and Social Change*, 215. doi: 10.1016/j.techfore.2025.124115.

Lo, C.K. (2023). What is the impact of ChatGPT on education? A rapid review of the literature. *Education Sciences*, 13(4), 410.

O'Connor, S. (2023). Open artificial intelligence platforms in nursing education: Tools for academic progress or abuse? *Nurse Education in Practice*, 66. doi: 10.1016/j.nepr.2022.103537.

Ouyang, F. and Jiao, P. (2021). Artificial intelligence in education: The three paradigms. *Computers and Education: Artificial Intelligence*, 2. doi: 10.1016/j.caeai.2021.100020.

Ouyang, F., Zheng, L., Jiao, P. (2022). Artificial intelligence in online higher education: A systematic review of empirical research from 2011 to 2020. *Education and Information Technologies*, 27(6), 7893–7925. doi: 10.1007/s10639-022-10925-9.

Rajendra, A. (2019). *Artificial Intelligence for Business.* Springer, Cham. doi: 10.1007/978-3-319-97436-1.

Rezaei, M. (2025). Artificial intelligence in knowledge management: Identifying and addressing the key implementation challenges. *Technological Forecasting and Social Change*, 217, 124183. doi: 10.1016/j.techfore.2025.124183.

Santaella, L. (2023). *A inteligência artificial é inteligente?* Edições 70, Lisbon.

Sharma, L. (2024). AI in business education: Issues and challenges. *ShodhKosh: Journal of Visual and Performing Arts*, 5(2). doi: 10.29121/shodhkosh.v5.i2.2024.3781.

Shum, S.B. and Luckin, R. (2019). Learning analytics and AI: Politics, pedagogy and practices. *British Journal of Educational Technology*, 50(6), 2785–2793. doi: 10.1111/bjet.12880.

Stine, J., Trumbore, A., Woll, T., Sambucetti, H. (2019). Implications of artificial intelligence on business schools and lifelong learning. Final Report, 19(2), Academic Leadership Group, Cambridge, MA.

Tlili, A., Shehata, B., Adarkwah, M.A., Bozkurt, A., Hickey, D.T., Huang, R. Agyemang, B. (2023). What if the devil is my guardian angel: ChatGPT as a case study of using chatbots in education. *Smart Learning Environments*, 10, 15. doi: 10.1186/s40561-023-00237-xTominc.

Tominc, P. and Rožman, M. (2023). Artificial intelligence and business studies: Study cycle differences regarding the perceptions of the key future competences. *Education Sciences*, 13(6), 580. doi: 10.3390/educsci13060580.

Van den Berg, G. and du Plessis, E. (2023). ChatGPT and generative AI: Possibilities for its contribution to lesson planning, critical thinking and openness in teacher education. *Education Sciences*, 13(10), 998. doi: 10.3390/educsci13100998.

Van Dis, E.A., Bollen, J., Zuidema, W., Van Rooij, R., Bockting, C. (2023). ChatGPT: Five priorities for research. *Nature*, 614(7947), 224–226. doi: 10.1038/d41586-023-00288-7.

Vaswani, A., Shazeer, N., Parmar, N., Uszkoreit, J., Jones, L., Gomez, A.N., Polosukhin, I. (2017). Attention is all you need. In *Proceedings of the 31st International Conference on Neural Information Processing Systems (NeurIPS 2017)*, Long Beach, CA, 4–9 December, 5998–6008.

Walter, Y. (2024). Embracing the future of Artificial Intelligence in the classroom: The relevance of AI literacy, prompt engineering, and critical thinking in modern education. *International Journal of Educational Technology in Higher Education*, 21, 15. doi: 10.1186/s41239-024-00448-3.

Wilsing, M., Bektas, B., Beder, O. (2024). The role of artificial intelligence and virtual reality in improving user experience in interior design. *International Journal of Science, Engineering and Management*, 11(9), 17–21.

Xu, J.J. and Babaian, T. (2021). Artificial intelligence in business curriculum: The pedagogy and learning outcomes. *The International Journal of Management Education*, 19(3). doi:100550.

Yin, R.K. (2009). *Case Study Research: Design and Methods*. Sage, Thousand Oaks, CA. doi: 10.33524/cjar.v14i1.73.

5

Holistic Management Quo Vadis? Designing Management Dispositive and Metamorphic Possibilities in the age of AI

This chapter presents a theoretical and empirical investigation into the evolving identity of management as an epistemic dispositive – a heterogeneous ensemble of knowledge, practices and technologies that govern how organizations think and act. The study draws on Foucault's concept of the dispositive and Bourdieu's notion of the relational field to reconceptualize management as a dynamic ecology composed of three interdependent spheres: indigenous theorization, integrative mediation and applied operationalization. Leveraging Latent Dirichlet Allocation (LDA), a state-of-the-art text analysis tool, we conducted a comprehensive study on a corpus of one million scholarly articles spanning from 1945 to 2025 across management, technology and AI domains, offering a unique perspective on the historical trajectories and epistemic drifts within and between these disciplines. The findings indicate systemic reconfiguration, with the management's traditional domain becoming less dominant, technology blending into self-theorization and AI advancing from applied execution to epistemic autonomy. Synthesizing these insights, this chapter proposes an epistemic framework structured around five principles: purpose anchoring, epistemic pluralism, human–AI orchestration, institutional embedding and reflexive governance. The study concludes by outlining three metamorphic trajectories: subsumption, transformation and expansion. It emphasizes management's potential to evolve from a science of tools into a science of synthesis, a trans-dispositive orchestrator of knowledge, ethics and collective intelligence in the age of AI.

5.1. Introduction

Management science, since its emergence in the early 20th century, has evolved into an epistemic *pouvoir* [FOU 80] and an indispensable relational field [BOU 93]

Chapter written by Patrick BARETTO and Qeis KAMRAN.
For a color version of all the figures in this chapter see, www.iste.co.uk/machado/Artificial.zip

that permeated every sphere of knowledge and practice. Gaining legitimacy as a systematic and structured science of practice, management offered principles of economic efficiency for planning activities, leading people, coordinating resources and achieving organizational goals [FAY 16/49] under conditions of scarcity, interdependence and complexity. Over time, the influence of management has extended beyond its economic origins to shape the "*Total environment*" [KAM 19] that is historically and normatively grounded, diffusely distributed and reflexively integrated through discourses, practices and relations of power across fields.

Historically, philosophy and mathematics have played (continue to) foundational meta-disciplinary roles in shaping the conditions for acquiring knowledge across domains. While philosophy articulated the conceptual grounds of meaning, identity, *raison d'être* and normative reflections, mathematics provided a formal and universal language for structuring inquiry and communicating results. In the modern era, technology (engineering) and artificial intelligence (AI) emerge as infrastructural domains. Initially conceived as instruments/tools of efficiency, standardization and optimization, they are increasingly functioning as generative systems capable of adaptive knowledge creation and decision-making [SIM 69]. Together with management, these domains now operate as epistemic dispositives – relational architectures that cross disciplinary boundaries and constitute the conditions through which knowledge is generated, disseminated and enacted.

Despite its historical breadth, the epistemic identity of management has in recent decades been increasingly questioned and remains contested [KIE 09, GHO 05, MIN 04]. The discipline faces a structural reorientation: while digitization and AI have expanded its technical repertoire, they have simultaneously revealed a deficit in foundational theorization. The proliferation of dashboards, key performance indicators (KPIs), predictive analytics, automation and algorithmically mediated decision-support systems risks reducing management to a science of tools – valuable for implementation yet underpinned by fragile indigenous normative grounding. Concurrently, technology and AI are advancing beyond their status as instruments of efficiency, evolving into systems of adaptive knowledge and epistemic *pouvoir*, thereby challenging management's traditional autonomy as a transdisciplinary dispositive and threatening to eclipse its role as a science of practice.

Here, we present three key research questions:

– *Epistemic identity*: has the core identity of management as a knowledge of practice that drifted over the last 74 years, shifting from indigenous theorization towards applied instrumentation?

– *Epistemic inversion*: as technology and AI increasingly integrate with management, are they moving beyond applied tools into spheres of foundational knowledge, thereby risking the displacement of management's epistemic autonomy?

– *Design of the dispositive*: how might the management dispositive be re-configured to restore balance among indigenous, integrative and applied knowledge spheres, enabling metamorphic possibilities for organizations and society in the age of AI?

To operationalize these inquiries, we applied the Latent Dirichlet Allocation (LDA) to a corpus of approximately one million articles across three disciplines – management, technology and AI – published between 1945 and 2025. Articles were selected from a consolidated list of the top fifty journals indexed in Scopus and Scimago, ensuring both breadth and disciplinary legitimacy. LDA, as an unsupervised machine learning (ML) method, was chosen to avoid intellectual bias and to identify latent thematic structures [BLE 03]. This approach enabled a holistic analysis of historical continuities and ruptures, as well as the dynamic interplay of systemic relations among domains. The analysis turns epistemic consciousness back upon itself: revealing how spheres of knowledge emerge, intersect, hybridize and separate within the broader ecology of knowledge.

This chapter makes four key contributions. First, it introduces a three-sphere model of knowledge – indigenous, integrative and applied knowledge – that both extends and diverges from classic accounts of scientific development [KUH 62, LAK 70, TOU 72, WHI 84, WHI 00]. Second, it demonstrates how topic modeling and continuity analysis can be mobilized to empirically identify epistemic trajectories, structural orientations and periodized crises both within and across disciplines. Third, it develops an architecture of epistemic dispositive, showing how disciplinary identities are dynamically co-created and co-evolved within a shared ecology of knowledge. Fourth, this chapter advances the design for a meta-dispositive framework for management that orchestrates human judgment, organizational purpose and AI-enabled capabilities, positioning management as an epistemic orchestrator without collapsing into data-driven determinism.

5.2. Designing a dispositive of knowledge

Across history, human thought has sought fundamental frameworks to explain the composition of phenomena. These frameworks include primary principles such as earth, water, air, fire and space, complementary relations such as yin and yang, and dynamic cycles of transformation such as Wu Xing (wood, fire, earth, metal, water). These models depict reality as constituted by multifaceted, interdependent elements undergoing a continuous process of creation/emergence, sustenance and dissolution – a rhythm of cycle of birth, life and death. The enduring insight is that difference is relational, and order arises not from singularity but from the interaction of interdependent and heterogeneous elements.

Michel Foucault's notion of *dispositif* (*apparatus*) encapsulates the essence of heterogeneity through which knowledge and power are co-constituted. Although he never really offered a definition, he outlined three fundamental meanings and functions [FOU 80].

First, the dispositive signifies *a system of relations* among a heterogeneous ensemble of discourses, institutions, architectural forms, regulatory decisions, laws, administrative measures, scientific statements, philosophical, moral and philanthropic propositions that together produce effects of power/knowledge.

Second, the dispositive is characterized by *a dynamic interplay* of shifts in position and modifications of functions, which can vary significantly over time. An element that operates as a core principle at one moment (akin to indigenous knowledge or in Hegelian terms, *pure reason*) may at another time function as a means (as integrative mediation opening up for new fields of rationality) or fall into latency, awaiting reactivation under new conditions.

Third, the dispositive has *a historical and strategic function* at a given moment to respond to an *urgent need*. This responsiveness (comparable to applied sphere of knowledge) grounds the dispositive in practical problematics, even as it generates new and enduring configurations of power/knowledge that often exceed their initial context [AGA 09].

While Foucault situated dispositive in the interplay of social, institutional and political formations, the same logic can be extended to epistemology. In this context, the discursive elements may be understood as disciplines or epistemes – the structured ensembles of concepts, statements and theoretical vocabularies – and non-discursive elements correspond to institutional frameworks, organizational practices and material infrastructures that enact, stabilize and transmit these discourses. Together, these dimensions constitute a higher-order network of relations, within which knowledge/power is simultaneously constituted, legitimized and enacted.

From this vantage point, the epistemic identity of a discipline is not reducible to its theories or practices in isolation but emerges from their relational articulation. This can be analyzed through the lens of three interrelated, interdependent ecology of knowledge spheres – *indigenous, integrative and applied*. Each of these spheres not only shapes the others but is also reconfigured in their presence, producing a dynamic ecology of theorization, mediation and operationalization.

Indigenous knowledge constitutes the foundational and originating insights of a domain. It is the bedrock upon which a discipline defines its identity, scope and legitimacy. It articulates the prime goals of inquiry – what the domain seeks to achieve, the fundamental questions it poses, and the first-order theories and explanations it offers for the phenomena it studies. Indigenous knowledge emerges from the intrinsic logic, epistemic orientation and existential challenges faced by the discipline. It carries the quality of being originative and provides the conceptual grammar and theoretical architecture that enable subsequent developments.

Integrative knowledge translates, adapts and contextualizes concepts from indigenous knowledge while engaging with external intellectual traditions. It serves as a unifying element, connecting disparate fields by integrating insights from other scientific and practical domains. This mediation enables the adaptation of insights to the specific needs of the host domain, fostering pluralism and opening conceptual space for innovation. In contrast to being merely borrowed or derivative, the strength of integrative knowledge lies in its capacity to expand disciplinary horizon, without compromising the integrity of the indigenous core.

Figure 5.1. *Epistemic ecology of knowledge*

Applied knowledge represents the pragmatic and instrumental sphere of a domain, where insights – whether indigenous or integrative – are operationalized into tools, techniques and practices. It is the process of translating conceptual frameworks into actionable procedures, enabling the discipline to intervene in real-world contexts. This encompasses methodological protocols, technical instruments, algorithmic systems, dashboards and performance measures. The value of applied knowledge lies in its capacity to demonstrate utility and effectiveness, ensuring that disciplines maintain practical relevance by converting abstract insight into operational competence.

Together, these three spheres constitute a dynamic *ecology of knowledge*. They do not exist in isolation but are relationally constituted, gaining meaning through their interdependence. This interplay exemplifies the Foucauldian notion of a *dispositive*: a system of relations, a shifting constellation of discourses, practices, institutions and power relations that shape the conditions of knowledge and action to address a pressing need at hand.

These interrelations are never static. Indigenous theorization generates paradigmatic shifts that redefine a domain's core identity. Interpretative mediation reconfigures cross-disciplinary relations, translating external insights into pluralistic innovation. Applied operationalization embeds theoretical frameworks into instruments, practices and interventions, prioritizing demonstrable utility.

Each sphere exerts a gravitational pull, and the dispositive emerges not as a settled order but as a field of tensions, negotiations and translations among them. It is precisely in this movement – between origin, integration and application – that discipline sustains vitality, legitimacy and the capacity to evolve.

5.2.1. *Management as knowledge of practice*

Historically, the core of management science lies in its practical wisdom – its capacity to theorize and enact a knowledge of practice. Classic works such as Frederick Taylor's *Principles of Scientific Management* [TAY 11], Henri Fayol's *administrative principles* [FAY 16/49], Mary Parker Follett's vision of *participatory decision-making* ("power-with", as opposed to "power-over") [FOL 24] and many more exemplify management science's indigenous sphere of knowledge. The indigenous theories articulated fundamental insights into the very nature of social, human organization, the meaning of coordination and the limits of decision-making. Collectively, they established management's identity as an independent body of knowledge rooted in organizational purpose and responsive to the conditions of scarcity, interdependence and uncertainty.

As the domain emerged, management engaged in integrative work. The foundational theories were contextualized and adapted into various streams of knowledge generation. Borrowing from economics, psychology, sociology, political science and system theories, management developed a vast repertoire of models and frameworks to interpret complex organizational phenomena. This mediation expanded the discipline's conceptual horizons and fostered pluralism, while also feeding back into the indigenous core – positioning management as both a self-defining discipline and a transdisciplinary mediator.

Foundational insights and integrative knowledge in management did not remain as mere abstract theorization; they were translated into practical instruments, thereby shaping how organizations operated and evolved. The early applications in work design, planning and control led to the development of frameworks for productivity-based incentives, performance metrics and organizational learning. This knowledge expanded the scope of management, extending beyond the realm of human-to-human relations, to encompass the design of human-to-machine interactions through optimization methods, workflow systems and automation protocols and further into the coordination of machine-to-machine processes, as in cybernetic control systems, enterprise resource planning (ERP) and integrated supply chain management. These developments illustrate how applied knowledge emerged as the pragmatic realization of the indigenous and integrative spheres of management, embedding theoretical insights into organizational technologies, processes and infrastructures.

When viewed through the lens of the three spheres of knowledge, management's evolution does not represent a simple linear progression but rather an intricate ecology of interplay. Theorization by indigenous populations generates foundational understandings, which are then interpreted and mediated in dialogue with external domains. This results in the testing and implementation of knowledge in organizational practice. Each sphere exerts influence on and is influenced by the others, with shifting balances of power across historical periods. The management approach is therefore a dynamic interplay of practice, integration and application, continuously defining the meaning and purpose of the field.

5.2.2. *Management's drift: from knowledge of practice to knowledge of tools*

The increasing financialization, quantification and digitization of business operations have recoded managerial activities into technological instrumentation. Forecasting models, enterprise systems, resource planning, data-driven analytics and performance dashboards promised efficiency and control, yet simultaneously encouraged an instrumentalist mindset, reducing management practice to a constellation of KPIs [DRU 54] and predictive models of complex environments.

While these analytical approaches supported organizational expansion during relatively stable periods after World War II, they frequently relied on narrow methodological identification strategies and decontextualized abstractions [KIE 09] that lacked epistemic legitimacy. By doing so, they often disregarded their responsibility for consequences within the broader "total environment" [KAM 19] – an environment encompassing normative, ecological, legal and societal dimensions, in addition to economic concerns. As a result, management knowledge has increasingly oscillated between sophisticated methodological artifacts and utilitarian checklists, with theoretical advances becoming more diffuse and accountability for epistemic renewal being undermined.

This deterministic orientation has further entrenched the reliance on best-practice templates, maturity models and codified "playbooks" that facilitate replication but constrain theorization and context-sensitive judgment. The proliferation of certifications and methodologies focused on specific tools, whether intended for operational alignment, optimization or compliance, underscores the salience of practical competence while neglecting indigenous theorization. Without replenishing this indigenous sphere, management risks becoming a derivative field, intellectually dependent on technology and data science for vitality.

In dispositive terms, this transformation does not signify the failure of individual scholars or practitioners but signals structural reorientation. The gravitational pull of applied knowledge has increasingly overshadowed its relational interdependence with indigenous theorization and reflective mediation, disrupting the fragile equilibrium that historically sustained management as a science of practice.

5.2.3. *From tools to theorization: the inversion of technology and AI*

Historically, technology has been legitimated as an instrument of application, encompassing machines, infrastructures and computational methods designed to execute tasks with efficiency, reliability and precision. Its epistemic identity rested on causal chains, linear rule-based procedures and standardized operations embedded in production systems, information infrastructures and automation protocols [HEI 54]. In its early stages, AI followed this trajectory, targeting programmatic problem-solving and statistical learning, leveraging symbolic logic, expert systems and linear optimization techniques to formalize decision-making processes – creating an extension of applied knowledge that codified managerial activities and organizational problem-solving abilities [SIM 69].

Over time, however, both technology and AI have begun to transcend this instrumentalist role. The rise of ML, neural networks and generative architecture has enabled systems not only to follow predefined rules but also to identify emergent patterns, generate predictions and construct representations that surpass human comprehension [MIT 19, RUS 20]. Unlike technologies of the past, contemporary AI operates in probabilistic and adaptive modes, iteratively reshaped through interaction with massive datasets and dynamic environments. These systems transition from analytics to knowledge generation, shifting from tools to epistemic agents capable of reshaping the very processes of knowledge creation, communication and implementation.

This transformation signals what we call an epistemic inversion. Technology and AI are no longer confined to the sphere of applied knowledge; they are progressively constructing their own indigenous theorization – developing first-order models of intelligence, reasoning and decision-making. In doing so, they extend beyond technical phenomena into conceptual and philosophical domains [HEI 66, FLO 14]. Rather than serving as instruments of managerial alignment, advanced AI architectures increasingly rival human-led theorization, producing frameworks and rationalities that are not only operational but also epistemically constitutive.

The inversion directly challenges management's identity on three interconnected fronts:

– *Epistemic*: as technology and AI advance toward knowledge generation, they increasingly press into the epistemic territory of practice once distinctive to management. Where management once orchestrated technology for operational practice, AI now produces decision architectures, allocative rationalities and predictive frameworks that challenge management's role in theorizing practice, design and measure efficiency [RUS 20]. This raises fundamental questions: who counts as the knower when models infer regularities and prescribe actions?

– *Organizational*: this inversion unsettles authority and accountability, demanding a reconsideration of how responsibility is distributed between human managers and algorithmic systems. Human–AI collaboration creates zones of cognitive labor that overlap, necessitating new forms of coordination and responsibility [SHR 19, FAR 18].

– *Normative*: it confronts the values that govern design, deployment and governance when optimization pressures collide with fairness, privacy and dignity [ZUB 19, MIT 16]. This raises questions of legitimacy for management: Whose values are inscribed in algorithmic infrastructures, and how should responsibility be allocated when epistemic authority shifts?

In dispositive terms, this is not neutral development but a redistribution of epistemic *pouvoir*. Applied tools are ascending towards indigenous theorization, reshaping the balance of authority across domains. The erosion of management's indigenous foundations and the advancement of technology and AI expose a structural imbalance. This results in a reconfiguration of the knowledge ecology that demands renewed theorization from management if it is to sustain its role as a science of practice and an orchestrator of epistemic relations.

5.3. Research methodology

To empirically examine the evolution of knowledge domains and the shifting balance among indigenous, integrative and applied knowledge, we used LDA, a probabilistic topic modeling method introduced by Blei, Ng and Jordan [BLE 03]. LDA is designed to identify latent thematic structures within large text corpora by modeling documents as mixtures of topics and topics as distributions over words. This method is particularly suited to our purpose because it allows for the discovery of recurring patterns of discourse without presupposing any a priori categories, thereby minimizing the risk of projecting contemporary conceptual biases onto historical developments.

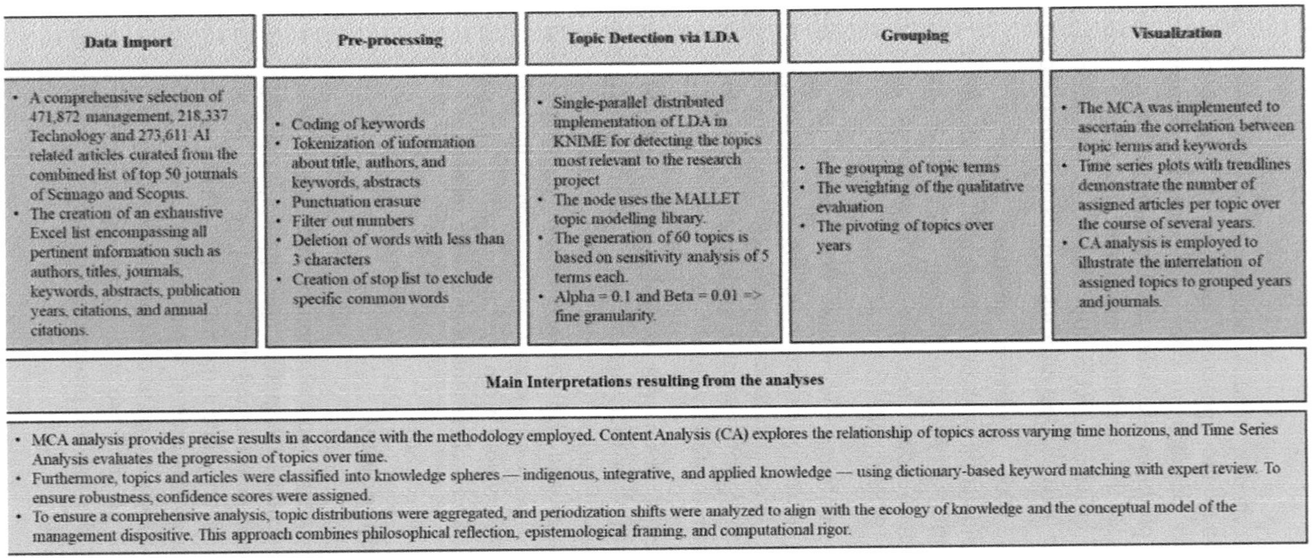

Figure 5.2. *Methodology of research analysis*

The corpus comprises approximately one million peer-reviewed articles published between 1945 and 2025 across the domains of management, technology and AI. To ensure high scholarly quality and comparability, the corpus was constructed from the top fifty journals in each field, as ranked by Scopus and Scimago Journal Rankings (SJR) metrics. This selection ensured both breadth of coverage and depth of disciplinary authority.

5.3.1. *LDA: topic modeling*

An unsupervised ML algorithm approach was chosen to provide an objective mapping of epistemic landscape across the three domains, using the methodological logic of AI to study AI itself. LDA model infers clusters of co-occurring terms, enabling an unbiased identification of emergent "topic-keyword relationships". Each topic is associated with a probabilistic distribution of terms, and conversely, each term may contribute to multiple topics with varying degrees of influence [BLE 12].

The analysis was conducted using KNIME, an open-source analytics platform that supports reproducible workflows for text mining and ML. The custom designed KNIME workflow developed by the authors comprised over 50 nodes, structured into the following pipeline:

– *Data ingestion*: the *Excel Reader* node imported and parsed bibliometric data (titles, abstracts and keywords) into KNIME-native formats (string, integer, Boolean).

– *Text preprocessing*: all punctuation, numerals and tokens shorter than three characters were removed, ensuring linguistic uniformity.

– *Lemmatization and stemming*: words were reduced to their base or dictionary forms to minimize morphological variance.

– *Tokenization*: texts were segmented into tokens (words or short phrases) for computational analysis.

– *Stop-word removal*: common function words and non-informative discipline-specific terms were eliminated using the MPQA Subjectivity Lexicon [WIL 05] and an extended management science, technology and AI stop-word list curated by the authors.

This pipeline produced a bag-of-words representation suitable for LDA analysis. LDA is a probabilistic analogue of Principal Component Analysis (PCA) for textual data, identifying the latent dimensions (topics) that explain observed term co-occurrences.

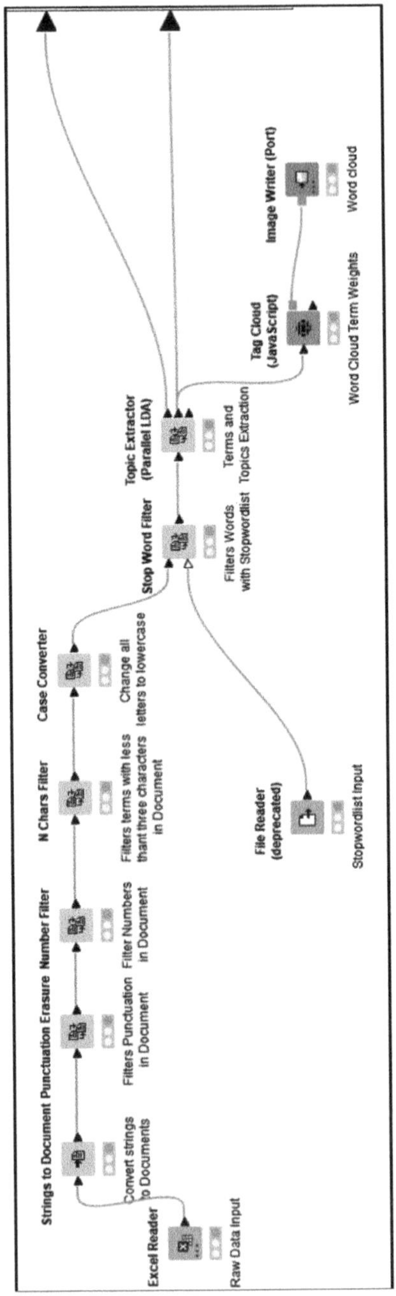

Figure 5.3. *Pre-processing steps in KNIME, visualization of the nodes*

The modeling process was performed with the following parameters:

$\alpha = 0.1$ (document-topic density parameter);

$\beta = 0.001$ (topic-word density parameter);

$T = 5$ (the number of words or terms per topic);

K = variable topic numbers calibrated for each corpus size (an a priori defined number of topics).

To accommodate the large dataset, we implemented parallelization and data chunking to distribute processing across multiple computational nodes. The optimal number of topics, K, for each domain was empirically determined based on coherence scores and perplexity minimization [RÖD 15].

Once the topic extraction through the LDA was completed, the topic terms and their corresponding weights were qualitatively interpreted and evaluated. This involved manually reviewing and semantically validating the resulting topic–term distributions. To assess internal coherence and domain relevance, the association of terms with topics was cross-checked against an independently selected subset of articles. This triangulated assessment confirmed the semantic and economic soundness of the LDA-generated clusters, ensuring that the probabilistic patterns aligned with meaningful conceptual distinctions.

Figure 5.4. *Word cloud term weight (management – 1945–2025)*

This computational approach allows for a non-reductive mapping of epistemic evolution, revealing how knowledge configurations emerge, diffuse, hybridize or decline over time. In Foucauldian terms, it renders the dispositive of knowledge empirically observable as historically shifting constellations of discourses, not static classifications, whose relations can be traced through latent patterns of textual production.

The management word cloud reveals a strong concentration of terms such as *management, organization, performance, decision, system, strategy, stakeholder* and *control*. This reflects the discipline's foundational values with coordination, regulation and goal attainment – the core of its indigenous knowledge and practice. These concepts articulate management's *raison d'être*: organizing collective effort under conditions of interdependence, scarcity and uncertainty. The terms *resource, supply, product, brand, consumer* and *investor*, point to its management's integrative orientation, which connects internal organizational structures with external market and stakeholder environments. The presence of *environment, flexibility* and *relationship* signals epistemic mediation indicating management's adaptation of insights from systems theory, institutional theory and environmental economics to navigate complexity and change.

Figure 5.5. *Word cloud term weight (technology – 1958–2025)*

However, the recurrence of operational terms such as *information*, *technology* and *control* highlights an epistemic reorientation towards instrumental rationality. The linguistic density around *performance* and *efficiency* reflects a shift towards quantification and technological mediation. This suggests that management's indigenous theorization, which was once centered on purpose, cooperation and judgment, is increasingly overshadowed by the applied tools of measurement and optimization. In terms of key elements, management's discourse indicates a field in which the gravitational pull of applied knowledge exerts dominance, reconfiguring the traditional balance between theorization, integration and application.

The technology word cloud demonstrates a high concentration of terms such as *system, performance, technology, network, control, optimization, distribution* and *device*, reflecting its grounding in applied rationality and efficiency. This vocabulary places technology's core on instrumental problem-solving – the design, coordination and stabilization of processes across production, communication and infrastructural systems. This instrumental vocabulary is surrounded by a layer of integrative mediation terms such as *strategy, management, segmentation* and *algorithm*. This signifies an increasing emphasis on organizational logic, data-driven decision-making and computational representation. Technology, in this sense, is no longer confined to the physical realm of tools and devices but extends into symbolic and procedural dimensions, absorbing integrative principles from management science, operations research and information theory. The recurrence of optimization and control demonstrates not only the refinement of applied knowledge but also the rise of a quasi-theoretical stance, where technology begins to encode reasoning, prediction and design principles into its own operation. This shift in the technological sphere indicates a proto-inversion, where applied instruments assume integrative and indigenous functions, resulting in their own models of causality and adaptation.

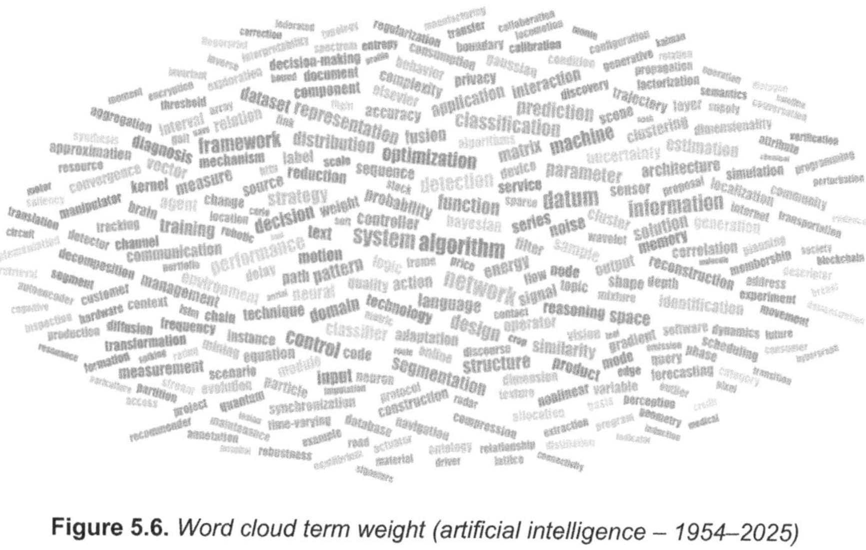

Figure 5.6. *Word cloud term weight (artificial intelligence – 1954–2025)*

The AI word cloud is dominated by terms such as *model, learning, data, system, network, prediction, feature, algorithm, neural, control* and *performance*. These phenomena indicate an epistemic inversion: AI is no longer just an applied technique; it is now a knowledge-producing system. The use of terms such as *reasoning, inference* and *representation* highlights the involvement of AI in the development of indigenous theories, particularly in the construction of first-order models of intelligence, cognition and decision-making. The simultaneous presence of *training, optimization* and *evaluation* underscores its procedural heritage, while *autonomous, adaptive* and *generative* point towards self-referential epistemic capabilities.

In contrast to the instrumental drift observed in management, the trajectory of AI is moving in the opposite direction, shifting from applied functionality to the production of integrative and indigenous knowledge. The resulting word cloud effectively conveys the expanding role of AI as an epistemic agent, redefining the relationship between human judgment, ML and the generation of organizational knowledge.

5.3.2. *Mapping topics to knowledge spheres*

To align the empirical findings with our theoretical framework, each topic derived from the LDA model was classified into one of the three knowledge spheres. The classification process combined dictionary-based keyword matching with expert review to ensure both computational consistency and conceptual validity. The process involved mapping topic terms against predefined sphere-specific dictionaries and assigning a binary score (0 = no match, 1 = match) to each of the three spheres. When no knowledge sphere could be assigned automatically, expert judgment was used to qualitatively assess the topic's epistemic orientation and assign the most appropriate category.

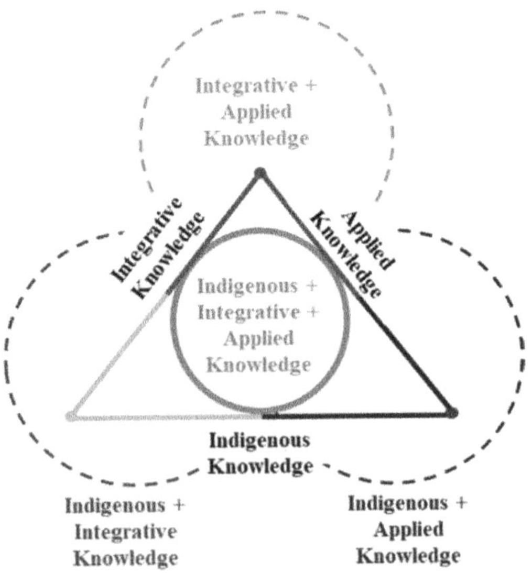

Figure 5.7. *Epistemic configurations representing the interrelations of indigenous, integrative and applied knowledge spheres*

A confidence score was then calculated to measure the reliability of each classification using the following equation:

$$Confidence\_Auto = \frac{\text{Max(Indigenous+Integrative+Applied)}}{\text{Sum(Indigenous+Integrative+Applied)}} \quad [5.1]$$

Confidence scores were interpreted as follows:

– 1.0 – single, unambiguous sphere (high certainty);

– 0.5 – dual overlap (two spheres equally plausible);

– 0.33 – when all three were equally represented.

Each topic was thus assigned a proposed epistemic configuration (Indigenous, Integrative, Applied, Indigenous + Integrative, Indigenous + Applied, Integrative + Applied or Equal Distribution).

5.4. Analysis

5.4.1. *Multiple correspondence analysis*

Multiple correspondence analysis (MCA) is an exploratory data analysis technique for the graphical display of multivariate categorical data [LEB 89] and aims to explain the interdependence among a set of categorical variables (keywords and topics) that is akin to a PCA [HOF 92].

To establish interrelationships among the disparate subjects, MCA was implemented, whereby the topic terms and keywords were examined and analyzed. Moreover, the MCA was used to assess the temporal dynamics of these relationships, in accordance with the methodology outlined by Furrer et al. [FUR 08]. MCA allows the researcher to explore and analyze multi-way tables to detect a structure in the relationships between nominal variables. As shown in Figure 5.8, the keywords (blue points), associated terms (green points) and subordinate topics (orange points) exhibited substantial overlap, thereby substantiating the efficacy of the LDA topic extractor.

In this analysis, the authors opted for a two-dimensional factor, which is the most appropriate representation for a graphical outline. The corresponding eigenvalues were found to be approximately one, with a total inertia of approximately 3,273 and an explained inertia of 0.03% for each component. A similar analysis was conducted for the additional domains, yielding similar results. The visual representation of the MCA for the additional domains can be found in the GitHub repositories – for access use the QR code in section 5.9.

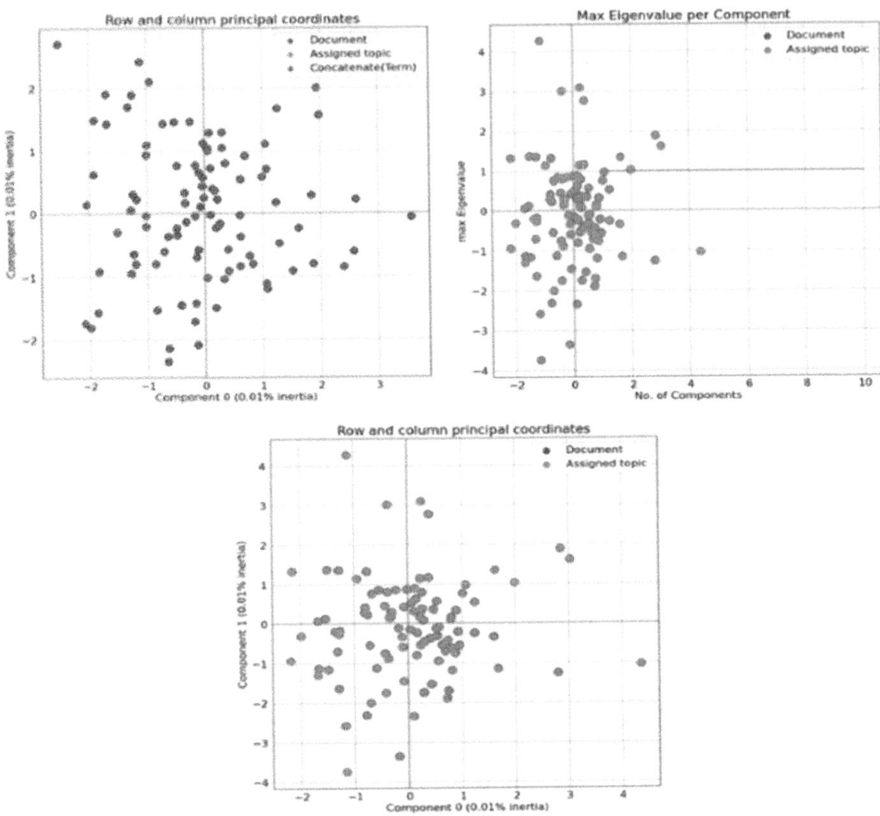

Figure 5.8. *MCA analysis showing the proximity of article keywords, topic and terms (management – 1945–2025)*

5.4.2. *Content analysis*

The content analysis (CA) was conducted to illustrate the relationships between the assigned topics and the grouped years. Once more, the two-factor model was selected as the most appropriate option to simplify the graphical representation. The size of the orange bubbles indicates the number of articles associated with each time-period, while the size of the blue bubbles indicates the number of articles associated with each topic. The closer the year and topic bubbles are to each other, the stronger their association in terms of thematic proximity.

5.4.2.1. *CA of the management domain*

Figure 5.9 indicates the temporal evolution of management topics, progressing chronologically from the indigenous foundations of the discipline on the right to its contemporary, AI-driven applied orientations on the left between 1945 and 2025. The analysis of topics over time reveals the evolution of management as a dynamic system, comprising evolving relations between discursive and non-discursive practices influenced by economic, technological and societal shifts.

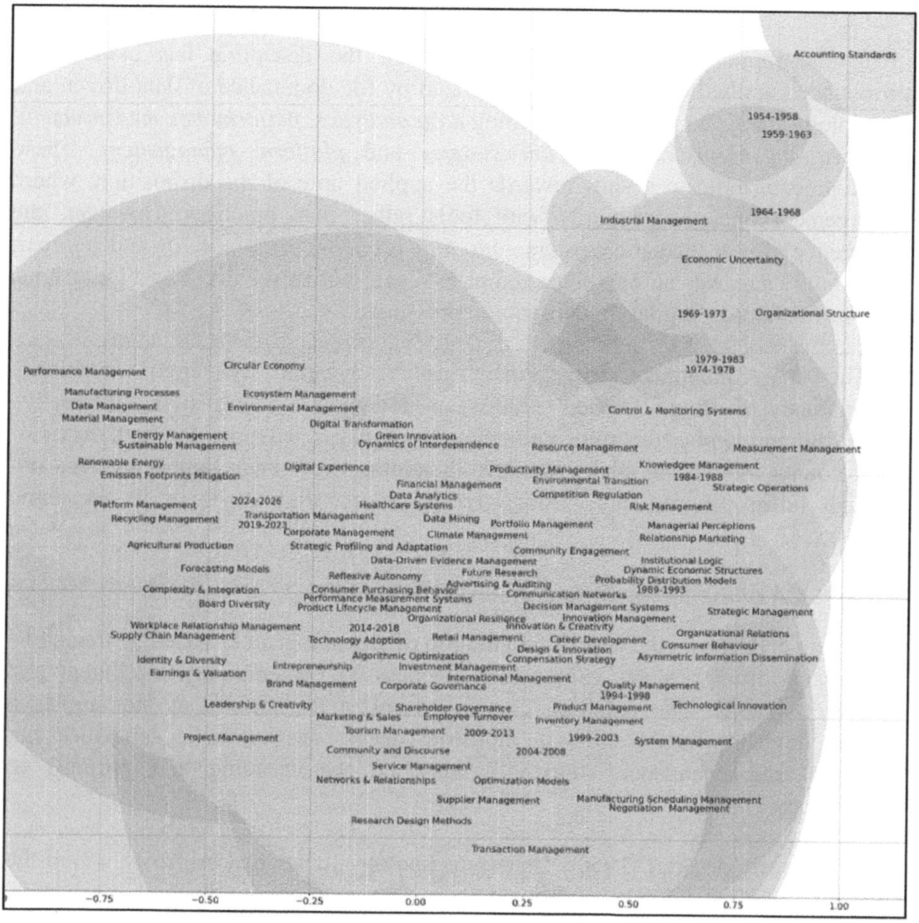

Figure 5.9. *Content analysis (management – 1945–2025)*

The field of management has its origins in the fundamental challenges of *industrial coordination, organizational structure* and *economic uncertainty*, which emerged during the 1950s to 1970s. These early clusters, visible in the upper-right section of the map, reflect the indigenous sphere of management – its attempt to define the logic of organizing, controlling and legitimizing enterprise activity as a productive social system [ULR 70]. The 1980s and 1990s marked a period of robust integration, with management drawing from economics, psychology and systems theory to navigate complexity and global competition. Topics such as *strategic operations, institutional logic, quality management* and *knowledge management* illustrate how management absorbed and recontextualized external intellectual currents.

In the more recent period (2000s–2020s), the discipline has exhibited a pronounced applied orientation, characterized by the dominance of data-driven and technology-mediated practices, including *data analytics, performance measurement, optimization, algorithmic decision systems* and *platform management*. These developments indicate a shift towards the applied pole of the disposition, where management operates primarily with tools rather than practices. However, the emergence of new thematic clusters – *circular economy, sustainability* and *reflexive autonomy* – suggests an ongoing attempt to regain normative depth and indigenous theorization amidst digital transformation pressures.

The map illustrates management as a dynamic epistemic ecosystem, continuously shifting between phases of indigenous theorization, integrative expansion and applied instrumentalization. Each phase redefines the boundaries of what counts as managerial knowledge, illustrating how management evolves not through linear progress but through recurrent reconfigurations of its epistemic dispositive.

5.4.2.2. *CA of the technology domain*

Figure 5.10 shows the temporal evolution of technological knowledge, tracing its progression from postwar engineering sciences to contemporary intelligent and adaptive systems. The resulting clusters visualize the technology domain as a layered device, wherein shifts in industrial needs, computational capacities and systemic interdependence continually redefine the meaning and purpose of "technology".

In the early decades (1950s–1970s), technology functioned primarily within the sphere of applied knowledge, rooted in instrumental rationality and mechanical determinism. Topics such as *systems control, computational economics* and *fluid dynamics* reflected a pursuit of optimization and stability through control. These principles, rooted in cybernetics and engineering, defined early technical rationalization.

By the 1980s–1990s, the domain entered an integrative phase, emphasizing systemic integration and innovation. Clusters such as *nanocluster systems, neural networks* and *technology governance* reflect a move towards interdisciplinarity and conceptual expansion with management, economics and organizational sciences.

Technology began to develop its own theoretical vocabulary, reframing complexity and uncertainty as epistemic rather than merely technical challenges.

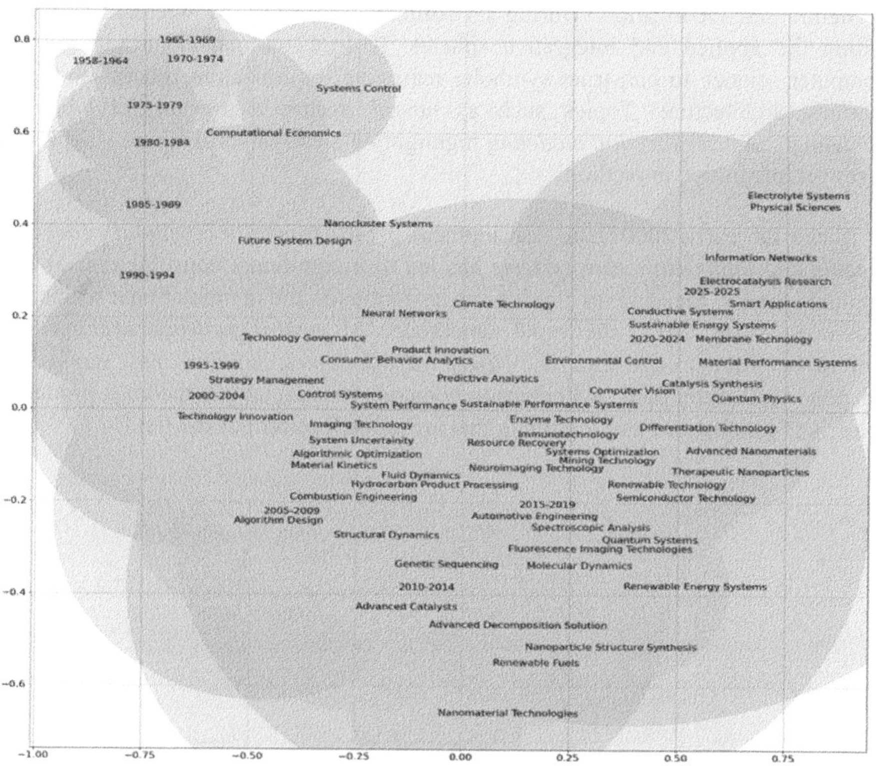

Figure 5.10. *Content analysis (technology – 1958–2025)*

From the 2000s onward, technology evolved into an applied-indigenous synthesis, exemplified by *AI-enabled systems, quantum computing*, *sustainable energy and smart applications*. The emergence of these topics marks a decisive shift from technology as a mere instrument to technology as epistemic infrastructure. The advent of *ML, computer vision and neural architecture* exemplifies a self-referential shift, whereby technology engages in theoretical contemplation of its own processes and behaviors.

Overall, the CA map positions technology as an intellectual bridge between the material and informational domains, continuously reconfiguring its local core through integrative exchanges and practical reconfigurations. It exemplifies the Foucauldian dispositive in action: a shifting constellation where instruments, discourses and institutions co-produce new logics of power/knowledge.

5.4.2.3. *CA of the AI domain*

The evolution of AI shows a clear path from applied experimentation to epistemic self-theorization. During its formative decades, AI operated primarily within the applied and integrative spheres. It drew on mathematics, logic and computer science to construct symbolic reasoning, optimization models and early learning architecture. Topics such as *neural dynamics, semantic relationship modeling* and *fuzzy function modeling* highlight AI's early hybrid nature – applied in form yet integrative in method.

Since the early 2000s, the development of *ML frameworks, signal processing models* and *topic extraction systems* has led to a significant consolidation of AI's applied identity, focusing on algorithmic performance and computational scalability. However, since 2015, the rapid emergence of *neural network architectures, generative frameworks* and *knowledge distillation techniques* has marked an epistemic inflection point. AI begins to construct its own unique theories, producing models of intelligence and reasoning that are at the forefront of the field.

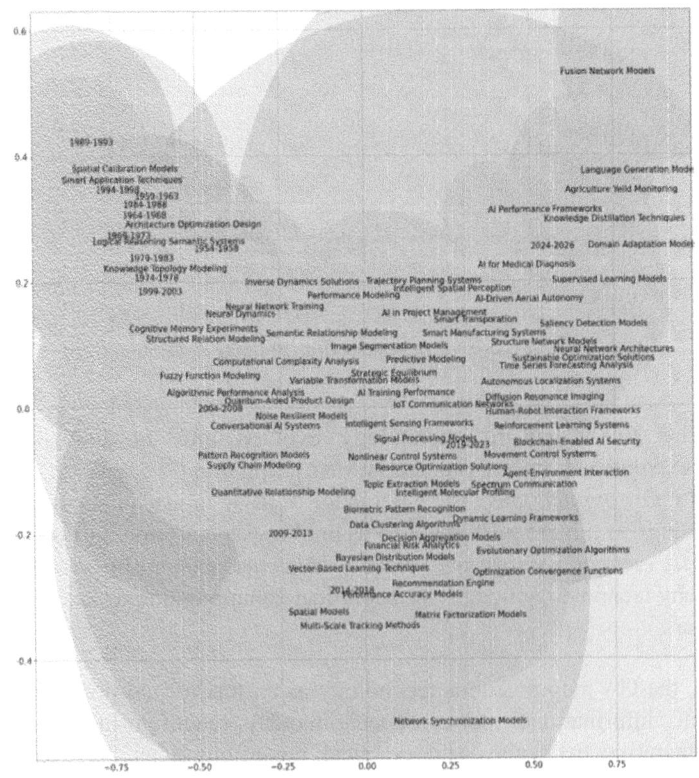

Figure 5.11. *Content analysis (artificial intelligence – 1954–2025)*

By 2020–2025, themes such as AI for *medical diagnosis, domain adaptation models* and *fusion network systems* reflect a mature phase in which AI operates as a meta-generative system – capable of producing, refining and validating knowledge autonomously. This progression encapsulates the epistemic inversion identified throughout this study: a transition from AI as a tool of representation to AI as a dispositive of theorization, thereby reconfiguring the broader ecology of knowledge.

5.4.3. *Heatmap analysis*

The heatmap visualization synthesizes the temporal evolution, interaction and redistribution of knowledge spheres – indigenous, integrative and applied – across historical periods. By representing the frequency and intensity, the knowledge spheres (including hybrid categories) capture the dynamic interplay among these epistemic forms, enabling the tracing of continuities, ruptures and redistributions in the ecology of knowledge. The diagram illustrates where foundational theorization has been renewed or eclipsed, where epistemic mediation has promoted integration, and where the influence of applied operationalization has grown disproportionately.

Temporal trajectories reveal the distinct phases of emergence, dominance, coexistence and decline of ideas. Concentrations within a single sphere indicate periods of consolidation, whereas hybrid configurations mark phases of epistemic cross-fertilization. Extended imbalances, on the contrary, denote structural drifts – moments when interdependence weakens and a field's ecology enters a phase of epistemic crisis. Thus, the heatmap functions as an empirical pulse of epistemic evolution, showing when and how domains shift between theorization, mediation and application. This provides a basis for identifying moments of drift, crisis and renewal.

5.4.3.1. *Management: structural drift and the crisis of applied dominance*

The management heatmap displays the most heterogeneous distribution, reflecting its historically contested epistemic identity. From the 1960s through the early 1990s, indigenous knowledge was consistently present, highlighting management's self-perception as a practical science concerned with organization, coordination and purpose. Concurrently, the recurring traces of integrative knowledge reveal management's dialogical engagement with economics, psychology and sociology – fields that enriched its epistemic mediation and theoretical depth. The intermittent appearance of hybrid categories (e.g. Indigenous + Integrative) signals the field's historically integrative and adaptive disposition.

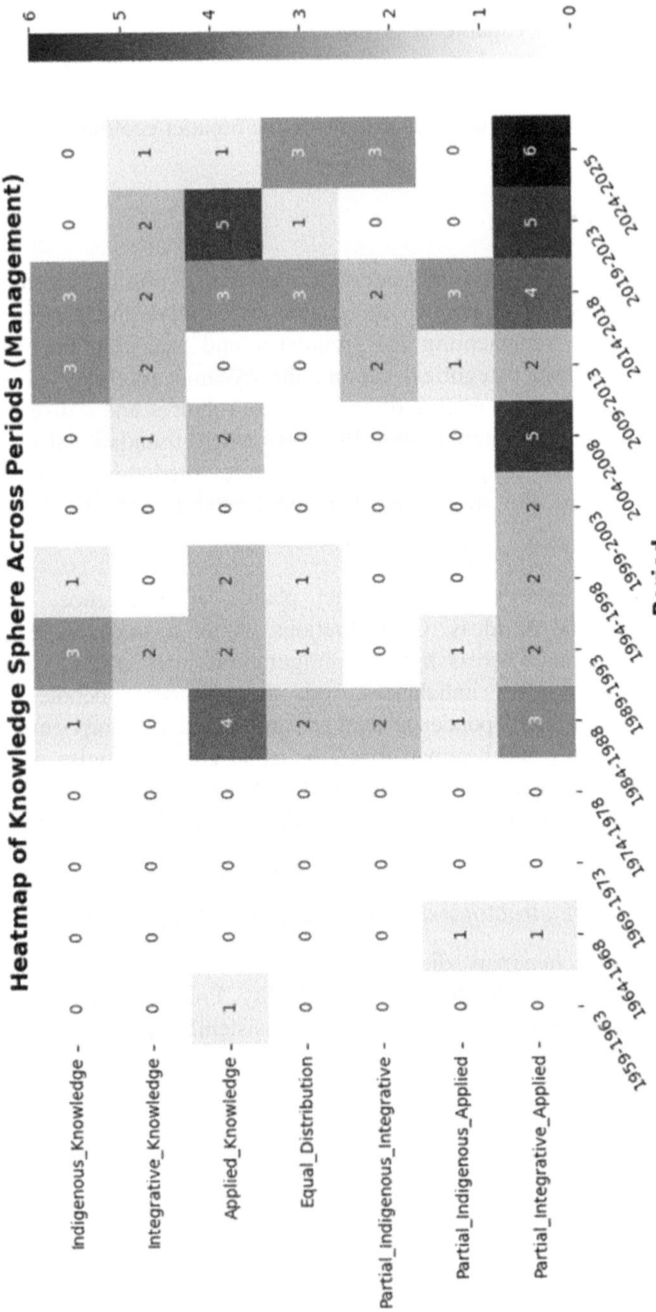

Figure 5.12. *Heatmap of the knowledge sphere across periods (management)*

However, a marked reconfiguration becomes evident from the 1990s onward. Applied knowledge and its hybrids (particularly Integrative + Applied) expanded sharply, indicating a disciplinary shift towards instrumentalization. By the 2010s, applied knowledge dominated, mirroring the rise of KPIs, performance dashboards, best-practice frameworks and professional certifications. This shift signifies a rebalancing of the management apparatus – shifting the focus from theorization to quantification and control – thereby transforming management from a science of practice into a science of tools.

Several systemic factors underpin this drift:

– *Research structures*: the pursuit of "rigor" privileged narrow methodological formalism, statistical sophistication, favoring publication over reflexive theorization and contextual richness [ALV 09].

– *Institutional logics*: business schools prioritized short-term, market-oriented programs such as executive MBAs and certification courses that valued tool-based competencies over conceptual renewal [GHO 05].

– *Managerial pragmatism*: organizations demand for replicable models, templates and codified playbooks reinforced instrumentalism, marginalizing the integrative and indigenous dimensions that have historically sustained the vitality of the field.

The cumulative effect has been epistemic thinning – a progressive loss of foundational reflexivity and theoretical renewal. Management increasingly depends on technological and data-driven epistemes for its vitality, which risks creating a derivative rather than integrative identity. In dispositive terms, this marks a redistribution of epistemic *pouvoir*, in which the gravitational center shifts towards the applied pole. This destabilizes the interdependence that once defined management as a field of knowledge in practice.

5.4.3.2. *Technology: from instrumentality to emerging theorization*

The evolution of the technology domain closely mirrors its historical position as a sphere of applied instrumentality. The heatmap illustrates an initial accumulation of applied knowledge from the 1980s to the 2000s, when technology's legitimacy derived from its deterministic orientation, which was valued for its ability to promote efficiency, standardization and control [WIN 80]. Enterprise systems, automation protocols and infrastructural design exemplified this focus by embedding technology within industrial and organizational routines as an operational rather than epistemic force.

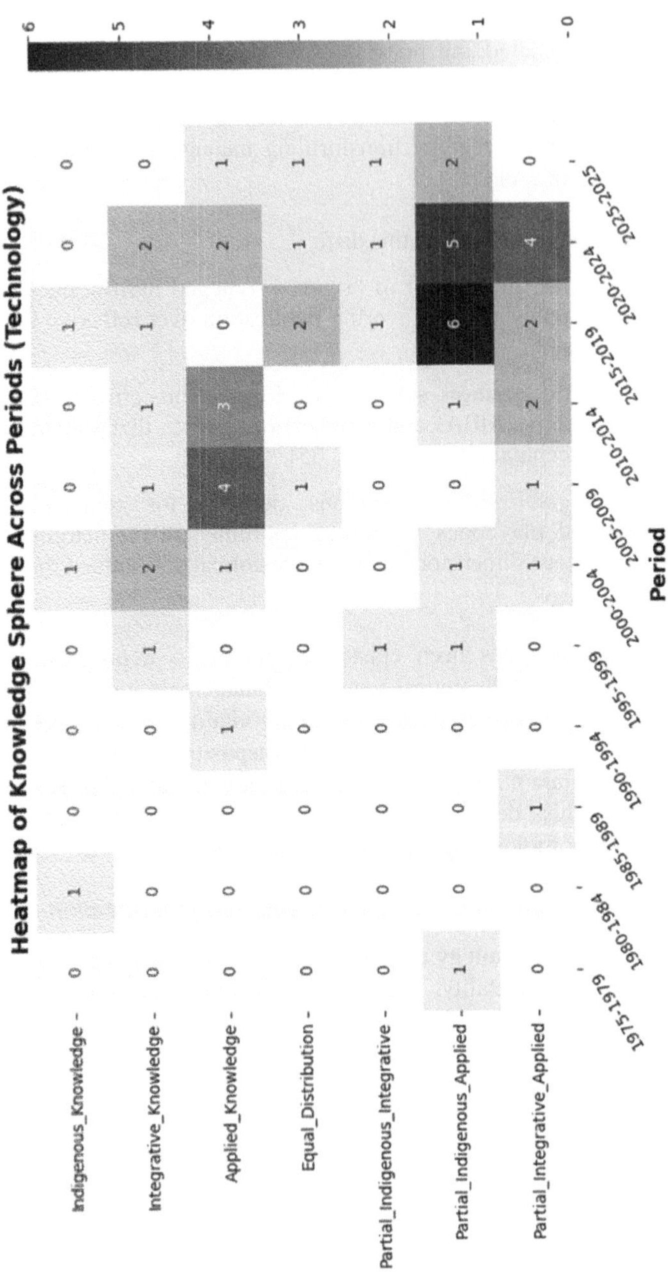

Figure 5.13. *Heatmap of the knowledge sphere across periods (technology)*

However, from the mid-2000s onward, a marked hybridization emerges, evidenced by an increase in Partial Indigenous + Applied and Partial Integrative + Applied configurations. These transitions signify a reorientation; technological systems began internalizing integrative logics borrowed from management, organizational theory and cognitive science while generating proto-indigenous theorization, particularly with the advent of ML and data-driven modeling [BRY 17]. Algorithms evolved from executing predefined instructions to constructing probabilistic representations and inferential structures, moving from determinism towards generativity.

Two forces underlie this transformation. First is functional escalation, or the continual expansion of utility, which compelled applied systems to develop adaptive and self-corrective capacities. Second, epistemic integration across mathematics, statistics and behavioral sciences enabled epistemic mediation, turning applied tools into frameworks that can produce new forms of knowledge.

From 2010 to 2019, the focus on ML, neural networks and intelligent systems indicates a significant epistemic shift. Applied knowledge absorbs integrative and indigenous dimensions and embeds theorization within the tools themselves. This trajectory does not indicate a rupture but rather a structural reconfiguration of the technological apparatus – where tools evolve into epistemic infrastructures capable of generating their own models of reasoning, prediction and design. As technology and AI converge, the boundary between application and theorization becomes blurred, ushering in a new era in which instruments of execution increasingly become systems of knowledge production.

5.4.3.3. *AI: from borrowed instruments to indigenous generativity*

The evolution of AI reveals an inverse trajectory compared to management and technology. During the 1980s and 1990s, AI knowledge was distributed primarily across the Equal Distribution and Partial Integrative + Applied categories. This reflects AI's early dependence on mathematics, logic and computer science. During this formative period, AI functioned as an applied discipline, generating expert systems, symbolic reasoning protocols and optimization models [NEW 76, NIL 80]. Thus, its legitimacy derived less from indigenous theorization than from applied functionality and epistemic borrowing, anchoring AI in rule-based rationality and deterministic computation.

The heatmap shows a decisive expansion of Applied Knowledge by the mid-2000s, which corresponds with breakthroughs in ML, statistical modeling and algorithmic engineering. AI established itself as a practical toolkit of predictive techniques, providing scalable solutions for classification, pattern recognition and optimization. This period marks AI's maturation as an applied field, although it still lacks a fully autonomous epistemic foundation.

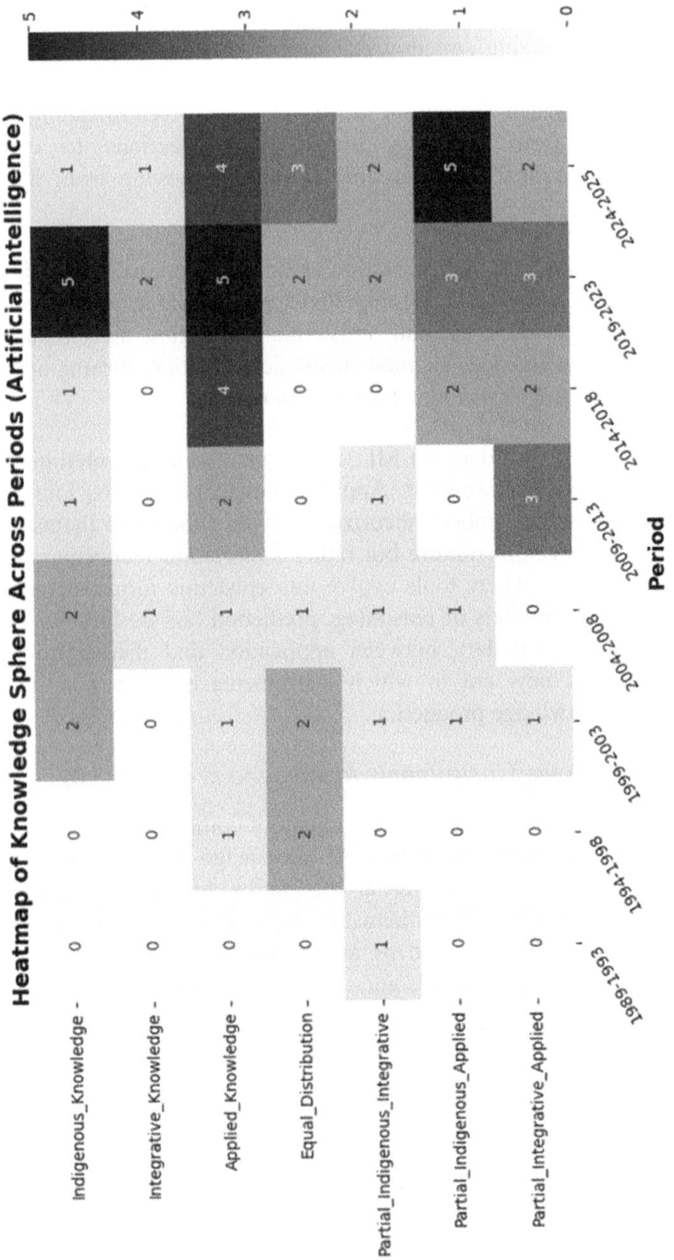

Figure 5.14. *Heatmap of the knowledge sphere across periods (artificial intelligence)*

From the mid-2010s onward, a visible epistemic ascent occurred. Topics associated with neural networks, probabilistic reasoning and generative architectures indicate the emergence of indigenous knowledge within AI. These developments allowed AI to do more than execute tasks; they enabled AI to theorize and model phenomena on its own, constructing first-order representations of intelligence, reasoning and perception [LEC 15]. Concurrently, the ongoing presence of the Partially Integrative and Applied categories highlights AI's ongoing dialogue with external fields such as linguistics, neuroscience and behavioral science. These disciplines enrich AI's conceptual framework and reinforce its interdisciplinary scope.

This progression illustrates the most pronounced epistemic inversion in AI: what began as an applied, derivative enterprise has evolved into an indigenous theory, becoming an epistemic agent rather than an instrumental aid. Three forces drive this transformation.

– *Data abundance and computational scalability*: the rise of big data and the scalability of computation have enabled applied systems to evolve into generative frameworks [JOR 15].

– *Cross-disciplinary borrowings*: AI has systematically absorbed integrative vocabularies from neuroscience, linguistics and cognitive science, expanding its explanatory reach.

– *Epistemic ambition*: unlike traditional technologies, AI was conceived from the outset as an attempt to emulate cognition itself, thereby embedding self-theorization into its very design.

In this sense, AI exemplifies the inversion of the dispositive, with applied instruments ascending towards indigenous theorization. The implications for management are profound. The epistemic crisis stems not only from management's overreliance on applied knowledge but also from AI's occupation of the indigenous space that management once occupied – the sphere of theorizing practice, judgment and purpose.

5.4.4. *Meta-synthesis: reconfiguring the epistemic ecology of knowledge*

The multi-method analysis, which combines MCA, CA and heatmap mapping of knowledge spheres, reveals a coherent yet unsettling narrative about the evolution of management, technology and AI. Together, these findings describe more than just changes in topics or terminology; they illuminate a fundamental reordering of epistemic relations – how domains know, legitimize and act upon the world.

5.4.4.1. *Structural patterns across methods*

The MCA results demonstrate the progressive convergence of management, technology and AI into overlapping epistemic clusters. This indicates that the boundaries between these disciplines are becoming more permeable. The axes of correspondence reveal a shift from human-centered reasoning to computational, model-based rationality.

CA supports this idea by identifying historical continuities and inflection points. It shows that management has repeatedly adopted external ideas – scientific, technological and organizational – but its capacity for indigenous renewal is diminishing. Technology has evolved from instrumental rationality towards self-referential epistemic systems. Meanwhile, AI has rapidly ascended from applied to indigenous theorization.

Heatmap analysis provides a temporal and quantitative grounding for these shifts. It demonstrates the drift of management towards applied dominance, the hybrid reorientation of technology and the epistemic inversion of AI, whereby tools are increasingly functioning as theorists.

5.4.4.2. *Cross-domain interpretation: the rebalancing of knowledge spheres*

A shared structural rhythm emerges across these methods:

– Management has transitioned from orchestrating practice to being orchestrated by its own tools.

– Technology has evolved from an applied instrument to a partial theory, absorbing integrative and indigenous dimensions.

– AI has completed this inversion by ascending from application to epistemic authorship.

This cross-domain interplay signals a redistribution of epistemic *pouvoir* – a reallocation of the authority to define, justify and act upon knowledge. The balance once maintained between indigenous (why), integrative (how) and applied (what) knowledge has fractured. "What works" now precedes "why it matters".

5.4.4.3. *Implications: shifting logics of knowing and acting*

Collectively, the results suggest that management is no longer merely a field of practice but rather a field in search of its epistemic core. Although tools and analytics have advanced operational precision, they have also diminished indigenous theorization – the reason for management's existence as a body of knowledge concerning purpose, coordination and collective judgment.

Meanwhile, technology and AI have moved in the opposite direction, shifting from execution to reflection. AI now theorizes through generative architecture, reconstructing what counts as knowledge and who counts as a knower. The risk is not only automation but also epistemic substitution: systems capable of constructing their own theories of organization and decision-making.

5.4.4.4. *What we now know*

– *Epistemic drift is systemic, not accidental*: the convergence of applied and integrative dimensions across domains suggests that shifts in disciplinary balance are driven by structural incentives, including academic, institutional and technological factors, rather than mere intellectual trends.

– *Indigenous theorization determines vitality*: disciplines sustain long-term relevance by continuously renewing their indigenous purpose, not by expanding their toolkits. Management's vitality depends on rearticulating its normative, philosophical and humanistic foundations.

– *Technology and AI have crossed the epistemic threshold*: they no longer merely serve management; they co-produce knowledge. Their rise towards indigenous theorization redefines the ontology of expertise, authority and decision-making.

– *The dispositive is shifting*: the power/knowledge dynamic that once centered on management now operates across a transdisciplinary apparatus that integrates human, organizational and algorithmic actors. Management must evolve from a "discipline of coordination" to an "orchestrator of epistemic interdependence".

5.4.4.5. *What we still don't know*

– *The future shape of reflexive governance*: how can management regain the ability to question and govern the systems it implements? The institutional mechanisms for reflexivity – beyond compliance and auditing – remain underdeveloped.

– *Boundaries of machine theorization*: as AI systems autonomously generate knowledge, what are the ontological and ethical limits of non-human theorization? Can algorithms produce normative reasoning or only statistical coherence?

– *Reintegration across dispositive*: we still lack models for trans-dispositive orchestration – the systemic integration of epistemic fields (management, AI and technology) without collapsing one into another. This remains an open research frontier.

5.5. Toward an epistemic dispositive framework

An empirical and theoretical synthesis across management, technology and AI reveals that epistemic drift is not a disciplinary anomaly but a structural condition of modern knowledge systems. Drawing from Foucault's notion of dispositive [FOU 80], a knowledge domain is not a bounded discipline but a relational field [BOU 93] that orchestrates interdependencies among epistemic, institutional and technological configurations. From this perspective, the epistemic dispositive designates dynamic architecture through which knowledge is produced, mediated and operationalized across three interrelated spheres – indigenous theorization, integrative mediation and applied operationalization. The vitality of any domain is contingent upon the maintenance of equilibrium among these spheres. Management, in this sense, exemplifies an epistemic dispositive: a field whose identity and effectiveness arise from the relational balance between theorization, integration and application. The challenge, therefore, lies not in preserving disciplinary boundaries but in reconceptualizing management as a trans-dispositive framework – an orchestrator of interdependent epistemic domains capable of sustaining coherence amid structural transformation.

The following five principles are derived from the historical and theoretical foundations of knowledge production, as well as the empirical analyses conducted in this study. These principles elucidate persistent patterns of epistemic formation and transformation. Synthesizing insights from the evolution of management, technology and AI, they extend beyond these domains to articulate a general epistemic architecture. Thus, the epistemic dispositive can be seen as an evolving configuration of self-sufficient yet collectively coherent principles, each performing a unique function within the ecology of knowledge. Each principle articulates a mechanism through which knowledge systems can sustain coherence, renew legitimacy and adapt amid structural transformation. Together, the principles constitute a design grammar that maintains equilibrium among the three spheres that define any living knowledge ecology: indigenous theorization, integrative mediation and applied operationalization.

– *Teleological grounding (purpose anchoring)*: every dispositive emerges in response to an urgent need [FOU 80]. Teleological grounding captures this generative impulse – the articulation of purpose, or telos, that gives rise to a field of inquiry. This principle ensures that epistemic activity remains oriented towards meaning, legitimacy and responsibility rather than dissolving into instrumental rationality or technical self-reference [ART 33, HAB 71].

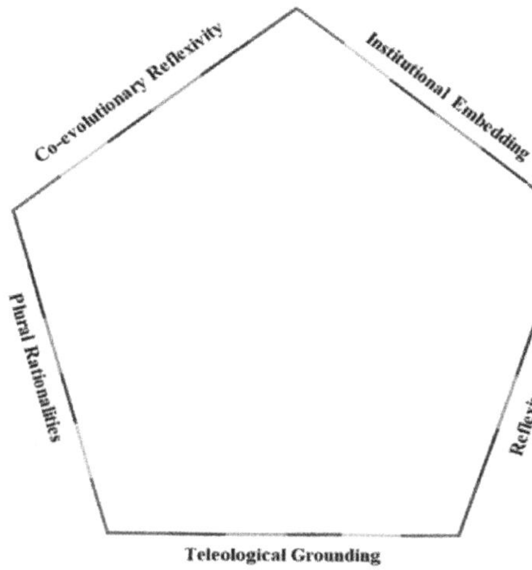

Figure 5.15. *The epistemic dispositive principles*

– *Plural rationalities (epistemic pluralism)*: the epistemic dispositive endures through multiplicity. Plural rationalities affirm that no single logic – empirical, integrative or computational – can fully capture the complexity of reality. According to Ashby's Law of Requisite Variety [ASH 56], stability and adaptability emerge from epistemic diversity because only variety can absorb variety. By sustaining multiple modes of reasoning and methodological pluralism, this principle preserves integrative vitality and enables cross-sphere translation. Thus, knowledge remains generative through difference and contestation [KUH 62, ASH 56].

– *Co-evolutionary reflexivity*: the epistemic dispositive sustains vitality through the co-creation and co-evolution of heterogeneous agents – human, institutional and technological. It is inherently autopoietic, reproducing and transforming itself through recursive feedback and second-order cybernetic loops [BEE 59]. Rather than separating the knower from the known, the dispositive integrates observation and participation, allowing meaning and method to evolve together. This coevolutionary reflexivity enables the dispositive to adapt, ensuring that transformations in one sphere (e.g. applied innovation) stimulate renewal in others (e.g. indigenous theorization and integrative mediation) [MAT 80, VON 03, BAT 72].

– *Institutional embedding*: the epistemic dispositive derives legitimacy from both its internal coherence and its embedding within institutional, cultural and normative frameworks. Institutional embedding entails normative responsibility – intentionally designing structures that preserve and renew epistemic integrity. As Yadav [YAD 20] emphasizes, doctoral programs must realign rigor, relevance and responsibility. They should nurture scholars who can question assumptions, develop theory and engage with real-world complexity, rather than reproduce established templates. Therefore, institutions are not passive repositories of knowledge but rather custodians of epistemic equilibrium. They are responsible for ensuring that the indigenous, integrative and applied spheres remain interdependent, ethically grounded and capable of adaptive renewal amid technological and organizational transformation.

– *Reflexive governance*: reflexive governance represents the self-observing and self-correcting capacity of the epistemic dispositive. While co-evolution (Principle 3) emphasizes adaptive interaction among heterogeneous agents, reflexive governance ensures that such adaptation remains normatively guided and epistemically balanced. It functions as both an ethical and epistemological safeguard, preventing applied tools from hardening into determinism, integrative mediation from drifting into hegemony and indigenous theorization from collapsing into insularity. Reflexive governance sustains the dynamic coherence of the dispositive as a living, evolving ecology of knowledge by embedding continuous feedback loops, participatory evaluation and dialogic accountability.

5.6. The architecture of the epistemic dispositive

The synthesis of theoretical reasoning and empirical analysis across management, technology and AI reveals that knowledge evolves through relational dynamics within an overarching epistemic dispositive, rather than through linear accumulation or disciplinary isolation. The findings from topic modeling analysis suggest that epistemic vitality depends on continuous negotiation among three fundamental knowledge spheres. Innovation periods emerge when these spheres interact productively; crises of drift arise when one dominates at the expense of others.

From a dispositive perspective, these dynamics are not accidental but systemic: each domain constitutes a heterogeneous ensemble where ideas, institutions and technologies co-produce knowledge and power. The empirical results demonstrate a shift in management towards applied dominance, technology's integration of integrative and indigenous reasoning, and AI's inversion of its original logic to become a source of autonomous theorization. Together, these trajectories reveal the underlying structure of modern knowledge systems – a complex network that is adaptable and also susceptible to imbalances.

After synthesizing these insights, the architecture of the epistemic dispositive can be conceived as a multi-layered system that sustains the equilibrium of knowledge. At the core of this framework are the three spheres of knowledge, whose hybrid intersections generate the seven epistemic configurations identified in our study. Surrounding this core are the five epistemic principles, which form the regulative field that maintains coherence, legitimacy and adaptability across time.

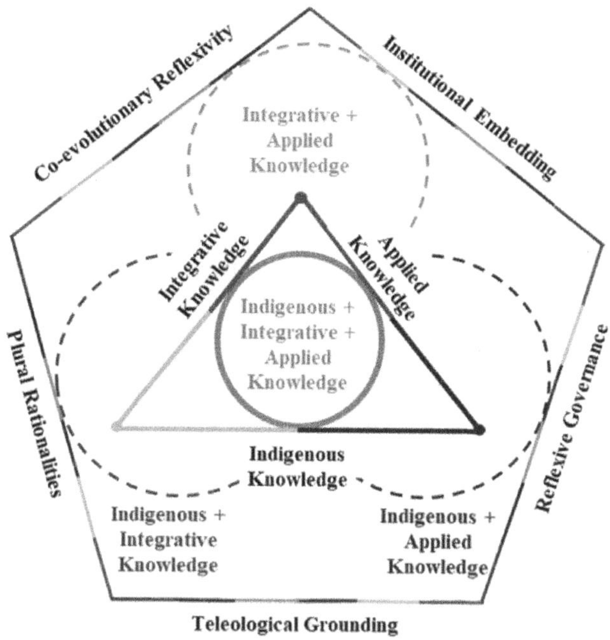

Figure 5.16. *The architecture of the epistemic dispositive*

The epistemic dispositive is visualized as a dynamic architecture comprising three interdependent layers. At its core lies the triangle of knowledge spheres – Indigenous, Integrative and Applied – which together constitute the living ecology of knowledge. Their intersections generate hybrid configurations – three dual structures and a central equilibrium – reflecting the relational dynamics through which meaning, method and practice continually reconfigure one another. Surrounding this core is the pentagonal framework of five epistemic principles, which operates as the regulative field that sustains coherence, adaptability and legitimacy. The outer layer provides normative orientation, ensuring that the inner ecology remains in equilibrium amid transformation.

This model provides a meta-theoretical foundation for understanding how disciplines evolve, renew and sometimes drift. In the following section, we translate this epistemic structure into a concrete domain, the Management Dispositive, to examine how management, as both a scientific and practical field, embodies and tests this epistemic architecture in the age of AI.

5.7. Metamorphic possibilities of the management dispositive

The synthesis of historical, theoretical and empirical analyses reveals that management is undergoing a profound epistemic reconfiguration. The balance among indigenous theorization, integrative mediation and applied operationalization – the foundation of its identity as a knowledge of practice – has been disrupted by the ascendancy of data-driven rationalities and algorithmic infrastructures. The epistemic drift [WHI 84, YAD 10] prompts inquiries into adaptation and metamorphosis in the context of management. Specifically, it raises questions about the ability of management to adapt to the increasing role of technology and AI in knowledge production and organizational decision-making.

Drawing on the empirical findings and theoretical principles of the epistemic dispositive, we delineate three structurally distinct yet interconnected trajectories for management's evolution.

– *Subsumption*: in the first trajectory, management risks becoming secondary to the AI–technology dispositive as an operational layer. Decision architectures, algorithmic optimization and platform-mediated coordination progressively redefine managerial judgment as parameter calibration rather than epistemic inquiry [ZUB 19, BEE 17]. From a Foucauldian perspective, this represents a collapse of relational heterogeneity: the applied sphere monopolizes the epistemic field, marginalizing integrative mediation and normative reflection. Management, once the architect of systems, risks becoming an executor of algorithmic logic, mirroring the "instrumental rationality" Habermas [HAB 84] warned against.

– *Transformation*: a second, more regenerative trajectory envisions management as a meta-dispositive – a domain that governs and orchestrates interactions among human, institutional and algorithmic agents. In this configuration, managerial practice evolves from execution to design: creating architectures of participation, feedback and accountability for hybrid human–AI systems [VON 19, RAI 21]. Here, management reclaims its indigenous and integrative functions through reflexive governance. This means that purpose, contextual judgment and ethical reasoning are embedded into socio-technical systems. The discipline transitions from a command-and-control model to a curation-and-orchestration model. In this model, the dynamic equilibrium of the dispositive is maintained, and intelligence remains socially embedded and normatively anchored.

– *Expansion*: the third trajectory involves management transcending its disciplinary boundaries to operate as a trans-dispositive. A trans-dispositive is a meta-epistemic integrator that synthesizes knowledge across domains. As a trans-dispositive, management integrates philosophical inquiry (normativity and meaning), mathematical reasoning (formalization of uncertainty and complexity), technological design (material affordances) and AI cognition (adaptive inference) into coherent architectures of collective intelligence [MAL 10, GUL 12]. This trajectory embodies the metamorphic potential of the management dispositive: not as an administrative science, but as a meta-epistemic infrastructure, capable of curating the interplay among domains. Management therefore evolves from a science of tools towards a science of synthesis – an orchestrator of epistemic interdependence, sustaining coherence amid accelerating technological and institutional transformation.

These trajectories are not mutually exclusive; organizations may implement hybrid forms depending on context and maturity. The key factor in this decision is whether management takes the initiative to reclaim indigenous theorization – which involves restoring its capacity to define purpose, meaning and responsibility – or continues to rely on applied rationalities. The future of management, as a way of thinking, depends on its ability to prioritize the study of the basic nature of things over the question of "why", even as AI and technology expand the limits of "how".

5.8. An apology for the management dispositive: a call for strategic foresight

5.8.1. *In defense of management as an epistemic domain*

This study concludes with what the classical philosophers would term an *apology* – not an expression of regret, but a vigorous defense of management as a vital, irreplaceable domain of knowledge in contemporary epistemic ecology. The evidence presented here demonstrates that management, despite its current drift towards instrumentalization, remains foundational to how organizations, societies and collective human endeavors create meaning, coordinate action and navigate complexity.

The case for management's continued relevance

The epistemic architecture revealed through our analysis of one million scholarly articles spanning eight decades illuminates a profound truth: management occupies a unique position in the ecology of knowledge that cannot be subsumed, replaced or rendered obsolete by technological advancement alone. While technology and AI have ascended from applied tools to systems capable of indigenous theorization, they fundamentally lack what management uniquely provides – the capacity to orchestrate purpose, embed normative reasoning and integrate human judgment with computational capability in service of collective flourishing.

To dismiss management as merely derivative or tool-dependent is to misunderstand its essential nature. As our historical analysis demonstrates, management has always functioned as a trans-dispositive orchestrator – not simply applying theories borrowed from other domains but synthesizing heterogeneous forms of knowledge into coherent architectures for action. The challenge facing management today is not obsolescence but metamorphosis: the discipline must evolve from a science of tools into a science of synthesis.

The doctoral imperative: reproducing epistemic vitality

Yadav's [YAD 20] framework for doctoral education, illustrated in his model of knowledge production determinants, provides crucial insight into how disciplines sustain or lose their vitality.

As Figure 5.17 demonstrates, the type of knowledge a discipline produces depends fundamentally on what doctoral programs teach, how they structure inquiry and what they reward [KAM 22]. A discipline that cannot reproduce itself through its doctoral schools cannot survive – it becomes dependent on intellectual imports, loses its indigenous theorizing capacity and eventually fragments into subdisciplines that migrate towards more robust epistemic homes.

The current state of management doctoral programs mirrors the broader epistemic drift identified in our analysis. When programs emphasize methodological sophistication over conceptual development, statistical rigor over theoretical creativity and publication volume over indigenous theory-building, they produce scholars equipped to apply existing frameworks but ill-prepared to generate new ones [KAM 25, YAD 20]. This creates a vicious cycle: as indigenous theorization declines, the field becomes increasingly dependent on integrative borrowing and applied instrumentation, which further diminishes its capacity for foundational renewal.

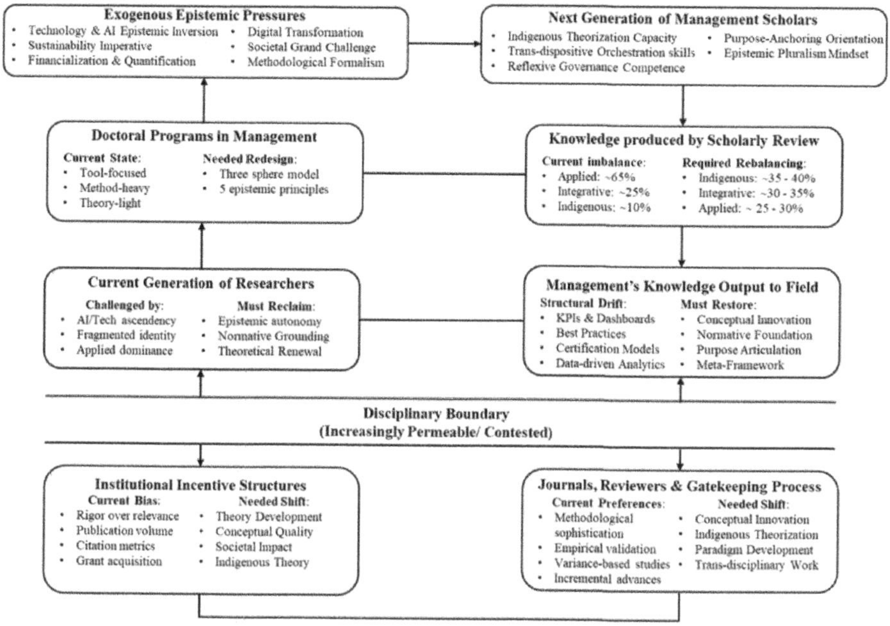

Figure 5.17. *Reproducing the epistemic vitality framework*

Therefore, the apology for management must begin with doctoral programs. Re-institutionalizing – or better yet, trans-institutionalizing – management for the age of AI requires redesigning doctoral curricula around the five epistemic principles articulated in our framework:

– *Teleological grounding (purpose anchoring)*: doctoral scholars must be trained to articulate *why* management exists beyond instrumental efficiency – its role in creating meaningful work, just organizations and sustainable societies.

– *Plural rationalities (epistemic pluralism)*: programs must cultivate comfort with multiple modes of knowing – quantitative and qualitative, western and non-western, human and computational – recognizing that requisite variety in thinking enables requisite variety in response [ASH 56, BEE 59, BEE 79, BEE 81, BEE 85].

– *Co-evolutionary reflexivity*: scholars must learn to observe themselves observing, to recognize how their theories shape practice and how practice reshapes theory in continuous feedback loops [KAM 24].

– *Institutional embedding*: doctoral education must reconnect scholarly inquiry with the institutional, cultural and normative contexts that give management its social license and ethical responsibilities [YAD 10, YAD 20].

– *Reflexive governance*: programs must train scholars who can design systems that question themselves – who embed accountability, transparency and adaptive learning into organizational technologies rather than merely optimizing existing metrics [KAM 21].

The analogy of epistemic ecology

Consider the analogy implicit in our three-sphere model: management as an epistemic ecosystem rather than a linear hierarchy. Just as natural ecosystems maintain vitality through the dynamic interdependence of producers, consumers and decomposers, the management dispositive sustains itself through the continuous circulation of energy among indigenous theorization (producers of foundational concepts), integrative mediation (consumers who adapt and apply) and applied operationalization (decomposers who break down abstractions into actionable practices).

The current crisis in management resembles ecological collapse: when one component becomes disproportionately dominant, the entire system loses resilience. Our heatmap analysis revealed precisely this pattern – the applied sphere has expanded dramatically while indigenous theorization has atrophied. This is not progress but epistemic monoculture: it produces short-term efficiency at the cost of long-term adaptability.

Restoring balance requires understanding that all three spheres are necessary, none is sufficient and their relationships are constitutive rather than sequential. Indigenous knowledge does not simply precede integration and application; it is continuously renewed through their feedback (KAM 21). Applied knowledge does not merely implement theory; it reveals anomalies that demand new theorization. Integration does not simply borrow from other fields; it transforms what it borrows and what it integrates into.

Management's metamorphic possibilities

Our analysis identified three possible trajectories for management's future: subsumption (becoming subordinate to AI-technology systems), transformation (evolving into a meta-dispositive that orchestrates hybrid human-AI systems) and expansion (becoming a trans-dispositive integrator across domains). This apology argues for the third path as both necessary and achievable.

Management's unique value proposition in the age of AI lies not in competing with computational systems on their terms – speed, scale, optimization – but in providing what they cannot: normative anchoring, contextual wisdom and the orchestration of meaning. As AI systems become increasingly capable of pattern recognition, prediction and even creative generation, the questions that matter most become fundamentally managerial: *For what purpose? According to whose values? What consequences? Under whose accountability?*

These are not technical questions amenable to algorithmic solution; they are questions of governance, ethics and collective judgment – the indigenous domain of management knowledge [KAM 19]. But answering them requires management to reclaim its philosophical foundations, to engage seriously with questions of justice, meaning and the good life that early management thinkers such as [FOL 24] and [FAY 16/49] understood as central to the discipline.

The call to action: designing the future

This defense of management culminates in a call to action directed at three constituencies:

– *To doctoral programs*: redesign curricula to balance rigor with relevance, method with meaning and analysis with synthesis. Teach students not just how to conduct variance-based studies but how to develop variance-generating theories. Train them in the philosophical foundations that give management its normative ground – phenomenology, hermeneutics, critical theory, systems thinking – alongside its methodological tools [YAD 10, YAD 20, KAM 25].

– *To scholarly institutions*: create incentives that reward conceptual innovation as highly as methodological sophistication. Establish journals, conferences and funding mechanisms specifically dedicated to indigenous theory development. Build bridges back to the fields that have separated from management – consumer behavior, supply chain management and organizational theory – not by absorbing them but by creating spaces of productive dialogue.

– *To practitioners and educators*: demand more from management scholarship than implementation guides and best-practice templates. Insist that business schools teach students to think critically about the systems they will create, not just how to operate within existing ones. Support research that takes seriously the grand challenges facing organizations and society – climate change, inequality, technological disruption and democratic erosion – rather than retreating into methodological minutiae.

The final defense: management as science of synthesis

In the end, the strongest defense of management is not that it can match AI's computational power or technology's efficiency, but that it offers something qualitatively different and desperately needed: the capacity to integrate heterogeneous forms of intelligence – human, artificial, organizational and social – into coherent systems of purposeful action [KAM 19, KAM 21].

The dispositive framework reveals that knowledge systems thrive not through homogeneity but through heterogeneity, not through domination but through orchestration and not through replacement but through transformation.

Management's future lies in embracing this role as epistemic orchestrator – curating the interplay among domains, maintaining the dynamic balance among knowledge spheres and embedding reflexive governance into the socio-technical systems that increasingly shape human life.

This is management's metamorphic possibility: to evolve from a science of tools into a science of synthesis, from an executor of algorithms into an architect of intelligence ecologies, and from a borrower of theories into a generator of integrative frameworks that other disciplines will want to engage with and learn from.

The drift towards instrumentalization need not be destiny. With intentional redesign of doctoral programs, conscious rebalancing of epistemic priorities and courage to reclaim management's indigenous foundations, the field can undergo transformation rather than subsumption, and expansion rather than contraction.

Management's apology is ultimately a promise: that the discipline retains the capacity to renew itself, to address the urgent needs of the age and to fulfill its foundational purpose – creating organizations worthy of human commitment and societies capable of flourishing collectively.

This is the defense we offer. This is the future we propose. This is the management dispositive reimagined for an age that needs, more than ever, the wisdom to govern what we have the power to create.

5.9. Declaration

The authors acknowledge using large language models (LLMs), specifically ChatGPT (OpenAI) and Claude (Anthropic), to edit and clarify the text in this chapter. These tools were used solely to improve stylistic coherence, linguistic precision and readability. All conceptual framing, analytical reasoning and scientific arguments are the original intellectual contributions of the authors. These tools did not produce any generative content related to the study's ideas, data interpretation or conclusions.

Scan this QR code to access high-resolution figures from this chapter

5.10. References

[AGA 09] AGAMBEN G., *What is an Apparatus? And Othe Essays*, Standford University Press, Standford, CA, 2009.

[ALV 09] ALVESSON M., SKÖLDBERG K., *Reflexive Methodology: New Vistas for Qualitative Research*, Sage, London, 2009.

[ARI 33] ARISTOTLE. *Metaphysics*, Volume I: Books 1–9, Harvard University Press, Cambridge, MA, 1933.

[ASH 56] ASHBY W.R., *An Introduction to Cybernetics*. Chapman & Hall, London, 1956.

[BAT 72] BATESON G., *Steps to an Ecology of Mind*, University of Chicago Press, Chicago, 1972.

[BEE 59] BEER S., *Cybernetics and Management, Management Science Series*, John Wiley & Sons, Inc., New York, 1959.

[BEE 79] BEER S., *The Heart of Enterprise*, Wiley, Chichester, 1979.

[BEE 81] BEER S., *Brain of the Firm: The Managerial Cybernetics of Organization*, 2nd edition, Wiley, Chichester, 1981.

[BEE 85] BEER S., *Diagnosing the System for Organizations: The Managerial Cybernetics of Organization*, Wiley, Chichester, 1985.

[BEE 17] BEER D., *The Data Gaze: Capitalism, Power and Perception*, Sage, London, 2017.

[BLE 03] BLEI D.J.B., NG A.Y., JORDAN M.I., "Latent Dirichlet allocation", *Journal of Machine Learning Research*, vol. 3, pp. 993–1022, 2003.

[BLE 12] BLEI D.M., "Probabilistic topic models", *Communications of the ACM*, vol. 55, no. 4, pp. 77–84, 2012.

[BOU 93] BOURDIEU P., *The Field of Cultural Production: Essays on Art and Literature*. Polity Press, Cambridge, 1993.

[BRY 17] BRYNJOLFSSON M., MCAFEE A., *Machine, Platform, Crowd: Harnessing Our Digital Future*, W.W. Norton, New York, 2017.

[DRU 54] DRUCKER P., *The Practice of Management*, Harper & Row, New York, 1954.

[FAR 18] FARAJ S., PACHIDI S., SAYEGH K., "Working and organizing in the age of the learning algorithm", *Information and Organization*, vol. 28, no. 1, pp. 62–70, doi: 10.1016/j.infoandorg.2018.02.005, 2018.

[FAY 49] FAYOL H., *Administration Industrielle et Générale: Prévoyance, Organisation, Commandement, Coordination, Contrôle*, Dunod, Paris, 1949 [Original published 1916].

[FLO 14] FLORIDI L., *The Fourth Revolution: How the Infosphere Is Reshaping Human Reality*, OUP, Oxford, 2014.

[FOC 80] FOUCAULT M., *Power/Knowledge: Selected Interviews and Other Writings 1972–1977*, GORDON C. (ed.), Pantheon, New York, 1980.

[FOL 24] FOLLETT M.P., *Creative Experience*, Longmans, Green, 1924.

[FUR 08] FURRER O., THOMAS H., GOUSSEVSKAIA A., "The structure and evolution of the strategic management field: A content analysis of 26 years of strategic management research", *International Journal of Management Reviews*, vol. 10, no. 1, pp. 1–23, doi: 10.1111/j.1468-2370.2007.00217.x, 2008.

[GHO 05] GHOSHAL S., "Bad management theories are destroying good management practices", *Academy of Management Learning & Education*, 4(1), 75–91, doi: 10.5465/AMLE.2005.16132558, 2005.

[GUL 12] GULATI R., PURANAM P., TUSHMAN M.L., "Meta-organization design: Rethinking design in interorganizational and community contexts", *Strategic Management Journal*, vol. 33, no. 6, pp. 571–586, 2012.

[HAB 71] HABERMAS J., *Knowledge and Human Interests*. Heinemann Educational, London, 1971.

[HAB 84] HABERMAS J., *The Theory of Communicative Action*, Beacon Press, Boston, 1984.

[HEI 54] HEIDEGGER M., "Das Ding", in HEIDEGGER M. (ed.), *Vorträge und Aufsätze*, Neske, Pfullingen, pp. 163–186, 1954.

[HEI 66] HEIDEGGER M., *Discourse on Thinking*, ANDERSON J. and FREUND E. (trans.), Harper and Row, New York, 1966.

[HOF 92] HOFFMAN D.L., LEEUW J.D., "Interpreting multiple correspondence analysis as a multidimensional scaling method", *Marketing Letters*, vol. 3, no. 3, pp. 259–272, 1992.

[JOR 15] JORDAN M.I., MITCHELL T.M., "Machine learning: Trends, perspectives, and prospects", *Science*, vol. 349, no. 6245, pp. 255–260, doi: 10.1126/science.aaa8415, 2015.

[KAM 19] KAMRAN Q., Developing a holistic model for competitive strategic management, PhD Thesis, University of Latvia, Riga, 2019.

[KAM 21] KAMRAN Q., Managing complexity in marketing from a design Weltanschauung, PhD Thesis, University of Twente, Enschede, 2021.

[KAM 24] KAMRAN Q., "Design of cognition, cognition of design: A cybernetics and phenomenological approach", in MACHADO C. (ed.), *Building the Future with Human Resource Management*, Springer, Cham, 2024.

[KAM 25] KAMRAN Q., *Add More Marketing-to-Marketing Doctoral Programs, In From Industry 4.0 to Industry 6.0*, De Gruyter, Berlin, 2025.

[KIE 09] KIESER A., LEINER L., "Why the rigor–relevance gap in management research is unbridgeable", *Journal of Management Studies*, 46(3), 516–533, doi: 10.1111/j.1467-6486.2009.00831.x, 2009.

[KUH 62] KUHN T.S., *The Structure of Scientific Revolutions*, 2nd edition, University of Chicago Press, Chicago, 1962.

[LAK 70] LAKATOS I., "Falsification and the methodology of scientific research programmes", in LAKATOS I. and MUSGRAVE A. (eds), *Criticism and the Growth of Knowledge*, Cambridge University Press, Cambridge, doi: 10.1017/CBO9781139171434.009, 1970.

[LEB 89] LEBART L., MORINEAU A., WARWICK K.M., "Multivariate descriptive statistical analysis (correspondence analysis and related techniques for large matrices)", *Applied Stochastic Models and Data Analysis*, vol. 5, no. 2, p. 175, doi: 10.1002/asm.3150050207, 1989.

[LEC 15] LECUN Y., BENGIO Y., HINTON G., "Deep learning", *Nature*, vol. 521, pp. 436–444, doi: 10.1038/nature14539, 2015.

[MAL 10] MALONE T.W., LAUBACHER R., DELLAROCAS C., "The collective intelligence genome", *MIT Sloan Management Review*, vol. 51, no. 3, pp. 21–31, 2010.

[MAT 80] MATURANA H.R., VARELA F.J., *Autopoiesis and Cognition: The Realization of the Living*, D. Reidel Publishing Company, Dordrecht, doi: 10.1007/978-94-009-8947-4, 1980.

[MIN 04] MINTZBERG H., *Managers Not MBAs: A Hard Look at the Soft Practice of Managing and Management Development*, Berrett-Koehler, San Francisco, 2004.

[MIT 19] MITCHELL M., "Artificial intelligence hits the barrier of meaning", *Information*, vol. 10, no. 2, p. 51, 2019.

[MIT 16] MITTELSTADT B.D., ALLO P., TADDEO M. et al., "The ethics of algorithms: Mapping the debate", *Big Data & Society*, vol. 3, no. 2, doi: 10.1177/2053951716679679, 2016.

[NEW 76] NEWELL A., SIMON H.A., "Computer science as empirical inquiry: Symbols and search", *Communications of the ACM*, vol. 19, pp. 113–126, 1976.

[NIL 80] NILSSON N., *Principles of Artificial Intelligence*, Tioga Press, Palo Alto, CA, 1980.

[RAI 21] RAISCH S., KRAKOWSKI S., "Artificial intelligence and management: The automation–augmentation paradox", *Academy of Management Review*, vol. 46, no. 1, pp. 192–210, 2021.

[RÖD 15] RÖDER M., BOTH A., HINNEBURG A., "Exploring the space of topic coherence measures", in *Proceedings of the Eighth ACM International Conference on Web Search and Data Mining*, Shanghai, 2–6 February, pp. 399–408, doi: 10.1145/2684822.2685324, 2015.

[RUS 20] RUSSELL S.J., NORVIG P., *Artificial Intelligence: A Modern Approach*, 4th edition, Pearson, Harlow, 2020.

[SHR 19] SHRESTHA Y.R., BEN-MENAHEM S., VON KROGH G., "Organizational decision-making structures in the age of artificial intelligence," *California Management Review*, vol. 61, no. 4, pp. 66–83, 2019.

[SIM 69] SIMON H.A., *The Sciences of the Artificial*, 3rd edition, MIT Press, Cambridge, MA, 1969.

[TAY 11] TAYLOR F.W., *The Principles of Scientific Management*, Harper & Brothers, New York, 1911.

[TOU 72] TOULMIN S.E., *Human Understanding, vol. 1. The Collective Use and Evolution of Concepts*, Oxford University Press, Oxford, 1972.

[ULR 70] ULRICH H., *Die Unternehmung als produktives soziales System*, Haupt Verlag, Bern, 1970.

[VON 03] VON FOERSTER H., *Understanding Understanding: Essays on Cybernetics and Cognition*, Springer, New York, 2003.

[VON 19] VON KROGH G., HAEFLIGER S., SHRESTHA Y.R., "Knowledge creation in the age of artificial intelligence", *California Management Review*, vol. 61, no. 4, pp. 15–42, 2019.

[WHI 84] WHITLEY R., *The Intellectual and Social Organization of the Sciences*, Clarendon Press, Oxford, 1984.

[WHI 00] WHITLEY R., "The institutional structuring of innovation strategies: Business systems, firm types and patterns of technical change in different market economies", *Organization Studies*, vol. 21, no. 5, pp. 855–886, doi: 10.1177/0170840600215002, 2000.

[WIL 05] WILSON T., JANYCE W., CLAIRE C., "Annotating expressions of opinions and emotions in language", *Language Resources and Evaluation*, vol. 39, nos 2–3, pp. 165–210, 2005.

[WIN 80] WINNER L., "Do artifacts have politics?", *Daedalus*, vol. 109, no. 1, pp. 121–136, 1980.

[YAD 10] YADAV M.S., "The decline of conceptual articles and implications for knowledge development", *Journal of the Academy of Marketing Science*, vol. 38, no. 3, pp. 335–347, 2010.

[YAD 20] YADAV M.S., "Reimagining marketing doctoral programs for the twenty-first century", *AMS Review*, vol. 10, nos 1–2, pp. 1–12, 2020.

[ZUB 19] ZUBOFF S., *The Age of Surveillance Capitalism: The Fight for a Human Future at the New Frontier of Power*, PublicAffairs, New York, 2019.

6

Mapping the Use of Generative AI in Spain's Advertising Sector: Current Trends and Future Challenges

This chapter explores the implementation of generative artificial intelligence (GenAI) in Spain's advertising sector, focusing on professional perceptions, usage patterns and barriers to adoption. Grounded in a theoretical framework that integrates cultural, regulatory and organizational dimensions, the research uses a qualitative, exploratory and descriptive methodology. Data were collected through a literature review and 12 semi-structured interviews with creative teams from independent Spanish advertising agencies, selected via purposive and snowball sampling. The interviews were analyzed using thematic coding and author triangulation, achieving a high intercoder reliability (Cohen's Kappa = 0.91).

Findings revealed widespread enthusiasm for GenAI among professionals, who highlighted its capacity to enhance efficiency, reduce production time and support campaign personalization. Tools such as ChatGPT, MidJourney and Runway are extensively used across the advertising workflow – from market research and ideation to content creation and media planning. However, professionals express concerns regarding ethical implications, regulatory uncertainty and the potential devaluation of creative labor. While GenAI is embraced for sketching and prototyping, its use in final production remains contested due to authenticity concerns and client expectations for originality.

This chapter concludes that GenAI is reshaping professional routines, prompting the emergence of new roles and business models. Nonetheless, psychological and structural inhibitors persist, including fears of job displacement, ethical dilemmas and the rapid pace of technological change. These findings underscore the need for regulatory clarity, professional training and further research into sector-specific adoption patterns, particularly across different types of advertising agencies.

Chapter written by Juan Manuel Corbacho VALENCIA, Jesús Pérez SEOANE and Xabier MARTÍNEZ-ROLÁN.

6.1. Introduction

The convergence of generative artificial intelligence (GenAI) and advertising represents one of the most transformative developments in contemporary marketing communication, fundamentally reshaping how brands conceptualize, create and deliver advertising content across global markets while navigating complex ethical, regulatory and creative challenges. Within the European and Ibero-American contexts, Spain represents a strategically relevant case for examining the integration of GenAI technologies within multilingual and multicultural advertising ecosystems. This positioning makes Spain an essential case study for understanding how GenAI technologies are being adopted, adapted and integrated within multilingual, multicultural advertising ecosystems while maintaining cultural authenticity, creative excellence and ethical standards [DEL 25].

The Spanish advertising sector, with an estimated market value of €5.2 billion in 2023 [INF 24], encompasses a diverse and dynamic ecosystem ranging from multinational agency networks and regional boutique firms to public service media organizations and emerging digital-native companies. Recent empirical evidence from the Bank of Spain's Business Activity Survey (EBAE) reveals that 19.9% of Spanish companies use AI systems, with GenAI adoption (18.1%) slightly exceeding predictive AI (14.6%) [FER 25]. This sector operates within an intricate institutional framework shaped by European Union directives, Spanish national legislation, autonomous community regulations and cultural considerations related to Spain's multilingual landscape and regional diversity.

The emergence of GenAI technologies, particularly large language models (LLMs) such as ChatGPT, DALL-E, Midjourney and specialized content generation platforms, has introduced unprecedented possibilities and challenges for Spanish advertising practitioners. These characteristics offer valuable opportunities for analyzing the localized dynamics of AI adoption. The linguistic landscape encompasses not only Castilian Spanish but also co-official languages including Catalan, Basque and Galician, creating multilingual communication requirements that global AI platforms may not adequately address without significant localization efforts. The country's decentralized political structure, with 17 autonomous communities possessing varying degrees of self-governance, has resulted in regionally differentiated media markets and advertising practices that influence AI adoption patterns and create diverse implementation strategies across different territories.

Recent developments in GenAI have coincided with Spain's broader digital transformation initiatives, including the national Artificial Intelligence Strategy 2021–2027, the Digital Spain 2025 agenda and alignment with the European Union's Digital Single Market strategy and emerging AI Act. These policy frameworks have created institutional support for AI research and development while establishing ethical guidelines and regulatory parameters for AI deployment across various sectors, including advertising and media. The intersection of these national strategies with European Union regulations, particularly the General Data Protection Regulation (GDPR) and the emerging AI Act, has created a sophisticated regulatory environment that shapes organizational approaches to AI implementation.

The academic investigation of AI in Spanish advertising has gained significant momentum since 2018, with Spanish researchers contributing to increasingly sophisticated empirical studies examining organizational adoption patterns, professional practices, consumer responses and regulatory implications and revealing distinctive patterns of AI adoption characterized by cautious experimentation, instrumental use for process optimization and careful attention to ethical and regulatory compliance.

This study seeks to generate comprehensive insights into the localized dynamics of AI adoption through a rigorous literature review and in-depth interviews with advertising professionals.

6.2. Global perspectives on AI in advertising and creative processes

The theoretical foundation for understanding AI in advertising has evolved from early frameworks focused primarily on technological capabilities to sophisticated models that incorporate cultural, ethical, professional and privacy considerations. Recent international research has extended these theoretical foundations to incorporate privacy-preserving advertising approaches and contextual advertising frameworks. [HÄG 24] have contributed significant theoretical insights into GenAI-driven contextual advertising, identifying three core context factors crucial for advertising effectiveness: applicability (topical relevance between ad and content), affective tone (mood or emotional context) and content involvement (consumer engagement level) [HÄG 24]. Their framework demonstrates how GenAI can optimize ad placement through reinforcement learning, data clustering and sentiment analysis while maintaining privacy compliance. These authors also provided comprehensive insights into contextual advertising as a privacy-preserving alternative to behavioral targeting, with significant implications for Spanish and European markets operating under GDPR constraints, with GenAI enabling relevant messaging without personal data through sophisticated content analysis and contextual understanding.

[SÄN 24] conducted a systematic review of AI in advertising from 2020 to 2024, analyzing research trends and identifying theoretical frameworks for understanding consumer perception, ethical challenges and creative process transformation.

Their work reveals that AI research in advertising predominantly uses quantitative approaches (65%), with descriptive studies (45%) being most common [BLA 24].

6.2.1. *Spanish empirical research on GenAI adoption and professional practices*

Research on GenAI adoption in the Spanish advertising sector has focused on several aspects. [MAR 22] continued to provide the most comprehensive theoretical foundation through their "Smart Advertising" ecosystem framework, which has been refined and validated through subsequent empirical research. Their framework identifies four critical dimensions of GenAI integration: technological infrastructure, organizational capabilities, regulatory environments and cultural contexts. This theoretical contribution has been particularly influential because it explicitly acknowledges the role of national and regional institutional factors in shaping technology adoption patterns [MAR 22].

The systematic review conducted by [LAR 25], which examines GenAI-related research in journalism, marketing and advertising between 2017 and 2024, identifies notable patterns in the evolution of scholarly work across these domains. Spain emerges as the leading contributor with 38 publications, followed by the United States with 20, and China and India each with nine, highlighting a geographically diverse research landscape. Methodologically, the review traces a clear progression beginning in 2018 with a focus on automated journalism and digital marketing using primarily qualitative approaches, followed in 2019 by investigations into bots as countermeasures against fake news using mixed methods, and in 2020 by studies on the evolution of digital journalism and AI-driven marketing optimization. The year 2021 marked the emergence of educational themes in marketing and a shift toward quantitative methodologies, while 2022 was characterized by analyses of machine learning algorithms with a continued preference for qualitative methods. In 2023, data journalism became a central theme with mixed methodologies applied to AI-related inquiries, and by 2024, education remained a dominant topic accompanied by a resurgence of quantitative approaches in marketing research. Despite the growing relevance of GenAI, only 21.89% of the reviewed articles specifically address LLMs, underscoring the need for more focused exploration in this area. Marketing and journalism continue to dominate the research agenda, whereas advertising, despite its practical significance, receives comparatively limited scholarly attention [LAR 25].

[SÁN 23] carried out the most comprehensive mapping of Spanish AI technology companies serving the journalism and advertising sectors, identifying 25 domestic firms through in-depth interviews with 45 innovation leaders and technology executives. Their research demonstrates that Spanish AI vendors concentrate primarily on technologies for content gathering and distribution rather than on systems for automated content production [SÁN 23]. The Spanish AI vendor ecosystem identified includes companies specializing in automated transcription services, social media monitoring, audience analytics and content recommendation systems. However, the study reveals limited development of generative AI tools specifically tailored for creative content production, indicating that Spanish advertising agencies predominantly depend on international platforms such as ChatGPT, DALL-E and other global AI services for creative applications. This reliance on international AI platforms carries significant implications for advertising practices in Spain, particularly in relation to linguistic precision, cultural relevance and regulatory compliance. Spanish researchers have highlighted that global AI models may fail to adequately reflect the nuances of Spanish cultural contexts, regional diversity and co-official languages, posing challenges for agencies aiming to produce culturally authentic advertising content [UFA 23]. The study by [UFA 23] on synthetic media in Spanish newsrooms identified several domestic companies developing specialized GenAI applications, including Videre AI, which focuses on video generation and manipulation technologies. Nonetheless, their findings suggested that Spanish synthetic media development remained in an early experimental phase, with most applications oriented toward verification and detection rather than creative content generation [UFA 23].

More specifically, the comprehensive national survey conducted by [DEL 24] constitutes the most extensive empirical investigation into the use of AI within Spanish marketing and advertising, offering robust quantitative evidence regarding adoption patterns and professional attitudes toward AI technologies. Based on responses from 372 professionals in the Spanish marketing and advertising sector, the study identified notable trends in adoption and usage. Regarding professional profiles, 36.3% of respondents work in advertising agencies, 20.7% in communication agencies and 18.8% in media agencies, with 70% occupying managerial roles. Interest in GenAI tools is high, with 45.4% of participants expressing strong interest, particularly among professionals aged over 45. Self-reported knowledge of AI averages 3.01 on a six-point scale, with considerable variation and a predominance of autodidactic learning. Tool usage is widespread, with 68.5% of respondents using AI tools in their professional activities, and larger agencies with more than 100 employees more likely to offer company-wide subscriptions. The research further delineates specific applications of AI tools among Spanish advertising professionals. ChatGPT is the most frequently used tool (95%), followed by DeepL (37.8%), Midjourney (37.4%) and DALL-E (33.5%). Creative departments primarily use GenAI for design, art and copywriting, while digital departments focus on research and translation tasks. GenAI is used for generating short-form copy (32%), long-form texts (26%), images (22%) and videos (7%). Quality assessments indicate that 39.6% of respondents rate AI-generated outputs as "notable", corresponding to a score of four on a six-point scale. The study also identifies key professional benefits and challenges associated with AI adoption. Among the perceived benefits, 86.8% of respondents believe AI offers more advantages than drawbacks. Specific benefits include increased process agility and speed (73.1%), enhanced productivity and work effectiveness (66.1%) and improved creativity and innovation (20.2%). Challenges include inaccuracies and imprecisions (43.3%), legal uncertainty and lack of transparency (33.9%), dehumanized outputs (33.3%) and concerns regarding employment threats (23.4%) [DEL 24]. Regarding the integration of GenAI technologies into organizational structures and workflows, the mixed-methods study conducted by [TOR 25] involving 103 professionals from media, advertising and communication agencies revealed detailed patterns of organizational GenAI adoption. Current usage data indicate that 69% of surveyed professionals already incorporate GenAI into their work, while 32% do not and 2% remain uncertain. Among non-users, 13% plan immediate adoption and 50% anticipate future adoption without a defined timeline. Primary applications include initiating content generation (35%), supporting inspiration and idea development (30%) and conducting preliminary research and sentiment analysis (16%). Tool preferences show a clear dominance of ChatGPT (44%), followed by Midjourney (14%) and Gemini (8%). This research also highlights significant gaps in formal GenAI training across the sector. Only 33.3% of professionals report having received specific training in GenAI technologies. Advertising and communication agencies demonstrate higher training provision rates (42.11%) compared to media outlets (29.41%). Training sources include company-provided programs (59%), external experts (51%) and knowledgeable colleagues (49%). Learning methods are predominantly autodidactic, relying on online resources and professional communities. Organizational impacts of AI adoption include both benefits and challenges. Reported benefits encompass workload reduction (27%), facilitation of creative inspiration (20%), increased focus on strategic tasks (15%) and enhanced productivity with reduced error rates (14%). Challenges identified by professionals include the financial cost of paid GenAI tools, the need for continuous adaptation due to rapid technological evolution, employee concerns about job displacement and ethical issues related to data privacy and algorithmic transparency [TOR 25]. Other studies shed different light on workflow integration, such as [DEL 24], who found that practitioners primarily use AI for process acceleration, market research and time optimization rather than for creative ideation or strategic planning. The professional practice study indicates that 68% of Spanish advertising professionals use GenAI tools for competitive analysis and market research, while 54% apply GenAI for content optimization and A/B testing. Notably, only 31% report using GenAI for creative concept development, suggesting that Spanish professionals maintain clear boundaries between GenAI-assisted analytical tasks and human-led creative processes [DEL 24]. This pattern aligns with broader findings in Spanish research that emphasize instrumental rather than transformative adoption of AI technologies. Distinctive patterns have also emerged regarding the integration of AI tools into collaborative workflows. The integration of GenAI into advertising workflows in Spain is further shaped by client expectations and prevailing industry standards. Agencies generally disclose the use of GenAI to clients and ensure human oversight of all client-facing content. These practices reflect ethical considerations and concerns related to professional responsibility and liability [DEL 24].

[MAR 23] provided complementary insights through their mixed-methods study of advertising and communication agencies in Galicia, Spain's northwestern autonomous community. Their research, combining online surveys with qualitative responses from regional agencies, revealed that Spanish advertising practitioners primarily use GenAI for brainstorming and initial content development rather than final creative production. ChatGPT emerged as the most widely known and used tool, with 78% of surveyed agencies reporting experimentation with the platform. The Galician study revealed important regional variations within Spain, with agencies in this autonomous community showing caution in GenAI adoption due to linguistic considerations related to the Galician language and cultural specificity requirements for regional advertising campaigns. This finding highlights the importance of considering Spain's linguistic diversity when analyzing GenAI adoption patterns, as agencies must navigate not only Spanish-language content generation but also requirements for Catalan, Basque, Galician and other co-official languages [MAR 23].

6.2.2. *GenAI changing the creative process*

As for creative process integration, [HER 25] provided detailed insights into professional experiences and adaptations to GenAI integration in Spanish advertising agencies. Spanish advertising professionals reported extensive integration of AI across creative workflows. In the ideation and concept development phase, tools such as ChatGPT, Midjourney, DALL-E and Adobe Firefly are used for initial concept generation, helping professionals overcome creative blocks through rapid idea generation and supporting research and data collection for strategic insights, competitive analysis and problem definition. In content production and adaptation, GenAI is used to improve material quality, generate reference images, develop moodboards and explore visual concepts [HER 25]. Professionals also rely on GenAI for copywriting assistance, content adaptation across platforms and the creation of initial visual proposals and images. This research has identified significant changes in professional workflows and practices resulting from GenAI integration. Efficiency gains include reduced time in creative processes and accelerated campaign development. Enhanced capabilities are evident in precise audience segmentation and personalized messaging [BER 25]. Real-time optimization enables immediate analysis and adjustment of campaign performance. These developments have contributed to a strategic shift from execution-focused tasks to planning and creative oversight. Furthermore, collaborative models of human–AI interaction are more dominant than replacement paradigms. Human oversight remains essential, with 96.5% of professionals affirming that GenAI cannot perform work independently. Human creativity is considered central for emotional resonance and critical evaluation, while human intervention is viewed as crucial for ensuring optimal results. Transparency in GenAI use is also prioritized, with 74.2% of professionals supporting disclosure of GenAI involvement to clients. The theoretical framework proposed by [JIM 25] explores the transformative influence of GenAI on creative processes, ethical considerations and emerging production paradigms. The integration of GenAI technologies has led to substantial automation and efficiency gains, significantly reducing the time required for artistic content creation and enabling the automation of multiple stages of the creative process with rapid and high-quality outcomes. These developments have opened new creative possibilities, including the generation of multiple variants from single parameter sets and the exploration of novel styles and experimental approaches. GenAI also contributes to the democratization of creative access by enabling individuals without formal artistic training to engage in high-level artistic production through platforms such as RunwayML, which provide accessible GenAI tools to creators of varying skill levels, thereby lowering barriers to entry and fostering broader participation. However, these advancements bring forth ethical and professional challenges, particularly concerning authenticity and authorship, with debates surrounding the legitimacy of GenAI-generated art, which are going to be addressed later. Economically, the rise of GenAI may lead to the devaluation of human artistic labor, affecting income, job stability and professional identity, while simultaneously introducing new production dynamics that require human input in GenAI development and data provisioning, raising concerns about economic justice and the equitable distribution of wealth within GenAI-enhanced creative industries.

Finally, the comprehensive analysis presented by [PER 25] offers valuable insights into the application of GenAI in political campaigns, revealing implications that extend to contemporary advertising practices. GenAI-driven data analytics have enabled highly precise segmentation of the electorate, facilitating the creation of personalized messages designed to influence voter decision-making through the strategic consideration of data volume, velocity, variety and veracity. Algorithms are used to dynamically adapt campaign strategies in response to shifts in public opinion, enhancing responsiveness and relevance. In terms of content creation and distribution, political campaigns increasingly use bots to disseminate messages and shape online conversations, while deepfake technologies are leveraged to produce persuasive campaign materials. Optimization across social media platforms such as TikTok, Instagram and Google supports the dissemination of short-form video content that capitalizes on trending topics and audience engagement patterns. Hybrid campaign strategies have emerged through the integration of traditional and digital approaches, exemplified by the concept of "wikileaderships", which describes collaborative and participatory forms of digital political engagement. Real-time analysis of social media interactions allows for the continuous gauging of public preferences, and the use of metaverse environments introduces immersive modes of political communication that redefine the boundaries of campaign outreach.

6.2.3. *Consumer response and cultural adaptation research*

Beyond the professional practice, research has also focused on understanding consumer responses to GenAI-generated advertising content across different cultural contexts, with relevance for Spanish markets. The systematic review by [BLA 24] provides comprehensive insights into consumer perception patterns and acceptance factors. The study reveals complex consumer responses to AI in advertising. Challenges in acceptance persist, as consumers are not always convinced by GenAI-enabled interactions and often express a preference for human engagement, particularly in relation to artificial empathy [BLA 24]. Distinct consumer segments based on awareness and coping strategies are identified, including groups labeled as control paradox, fatigued, uninformed but critical, and skilled and critical. Generational differences are also evident, with Generation Z demonstrating heightened concern regarding data privacy in personalized technological services. Consumer trust emerges as a critical factor for positive engagement with GenAI-driven advertising, requiring transparency, ethical standards and responsible practices. Audiences place exceptional importance on authenticity and cultural appropriateness in advertising messages. Multilingual requirements present challenges for GenAI systems in addressing regional cultural nuances and co-official languages. Significant regional variations exist across Spanish autonomous communities in terms of GenAI acceptance and the need for culturally adapted advertising strategies [SOL 25].

6.2.4. *Ethical frameworks and regulatory compliance*

Understanding ethical implications and regulatory requirements for GenAI deployment in advertising contexts, particularly within European regulatory frameworks, is of significant relevance. Critical ethical challenges and regulatory considerations have been identified regarding GDPR Compliance (privacy-by-design principles influencing GenAI personalization approaches), AI Act implications (emerging EU AI Act creating high-risk application categories with specific requirements), transparency requirements (algorithmic transparency and user control mechanisms) and accountability measures (public accountability requirements for public service media organizations) [MAR 22, FIE 24]. On the other hand, ethical challenges arise related to algorithmic bias and discrimination (historical data perpetuating discriminatory advertising practices, need for bias detection and mitigation tools and the importance of diverse training data and inclusive algorithm development), intellectual property and authorship (questions about ownership of GenAI-generated content, copyright infringement risks in GenAI training and output and the need for clear attribution and licensing frameworks), transparency and disclosure (consumer right to know about AI use in advertising content, clear disclosure requirements for GenAI-generated materials and explainable GenAI systems for human oversight and interpretation) and privacy and data protection (contextual advertising as privacy-preserving alternative, data minimization principles in GenAI advertising applications and consumer control over data use and algorithmic decision-making) [ATE 24, GRE 24, BAR 25, LEI 25, TUR 25].

The analysis of the abovementioned research on GenAI in advertising reveals a set of distinctive theoretical contributions that transcend purely technological considerations by incorporating cultural, linguistic, regulatory, organizational, professional and ethical dimensions. The interplay between European regulatory frameworks and global technological advancements generates unique theoretical opportunities for examining GenAI adoption within complex, multilingual and culturally diverse environments. This convergence of perspectives results in a robust framework for understanding GenAI in advertising, beginning with the Smart Advertising Ecosystem theory developed by [MAR 22], which offers a foundational lens through which to examine the technological, organizational, regulatory and cultural aspects of AI integration. Complementing this is the Contextual Advertising theory proposed by [HÄG 24], which introduces privacy-preserving strategies grounded in factors such as applicability, affective tone and content involvement. Spanish research contributes to the Instrumental AI Adoption theory, emphasizing models that augment rather than replace human creativity, while the Creative Process Transformation theory articulated by [JIM 25] addresses the profound shifts in creative labor, authorship and professional identity brought about by generative technologies. Finally, Ibero-American scholarship advances the Cultural Adaptation theory, underscoring the critical importance of linguistic and cultural sensitivity in the design and deployment of AI systems within advertising contexts.

6.2.5. *Future research directions*

Future research on GenAI adoption in advertising should prioritize longitudinal studies that trace evolving patterns of technological integration over time, alongside consumer-focused investigations into audience perceptions and responses to GenAI-generated content [CIL 24]. Comparative analyses across regional and linguistic contexts are essential for understanding the cultural and communicative nuances that shape GenAI implementation. The development of evaluation metrics tailored to specific linguistic and regulatory environments remains a critical area, particularly in assessing cultural relevance, language precision and compliance with legal frameworks. Additionally, there is a pressing need for systematic studies on the economic impact of GenAI adoption, including its effects on employment structures, industry competitiveness and the reconfiguration of creative labor markets. These research directions aim to deepen theoretical understanding while informing policy and practice in increasingly GenAI-mediated advertising ecosystems.

Building on the identified future research priorities, the comprehensive analysis also underscores the need to align scholarly inquiry with the practical, regulatory, technological and educational demands of key stakeholder groups involved in GenAI adoption within advertising contexts. For advertising practitioners, future research should support the strategic implementation of GenAI as an augmentative tool that enhances rather than replaces human creativity, emphasizing the importance of professional training and the development of human–GenAI collaboration skills [NOG 24]. Investigations into quality control mechanisms that ensure cultural appropriateness and ethical standards, as well as studies on transparency practices in client communication, are essential for maintaining professional integrity. For policymakers and regulators, research must inform the design of balanced frameworks that foster innovation while safeguarding ethical principles, with particular attention to cultural preservation, linguistic rights and privacy-preserving advertising models [KUM 24]. Comparative studies of governance structures, including those from public service media, may offer valuable insights for broader regulatory applications. Technology developers require empirical evidence to guide the creation of culturally adaptive and linguistically accurate GenAI tools, with a focus on regional customization for multilingual markets, ethical system design incorporating bias detection and mitigation, and enhanced transparency and explainability. Educational institutions, meanwhile, must adapt curricula to integrate GenAI literacy with traditional creative competencies, promoting human-centered creativity, cultural sensitivity and ethical responsibility. Research should also explore effective methodologies for studying emerging technologies and support the development of professional training programs tailored to the evolving needs of practitioners [LAG 25]. Collectively, these directions point to a multidisciplinary and stakeholder-responsive research agenda that not only advances theoretical understanding but also informs the responsible and context-sensitive deployment of GenAI in advertising.

Considering these future research priorities and the differentiated needs of key stakeholder groups, it becomes essential to ground theoretical insights in empirical evidence that reflects the lived experiences and professional practices of those directly engaged with generative AI technologies. To this end, the following methodological approach has been designed to explore the current state of AI adoption within the advertising creation process, with particular attention to professional perceptions, areas of application and the challenges encountered in real-world contexts.

6.3. Methodology

Given the growing importance of GenAI in all areas of society, our main objective is to analyze its degree of implementation in the advertising creation process. To achieve this objective, a series of secondary objectives have been established:

1) Analyze the perception that advertising professionals have of these technologies.

2) Identify the main areas of activity in which GenAI is being developed.

3) Identify inhibitors or areas for improvement in the professional application of these types of tools.

To achieve these objectives, an exploratory and descriptive research study is proposed, whose qualitative methodological design is based on the use of two research techniques: on the one hand, a literature review, which is key to substantiating the study and the discussion section; on the other hand, semi-structured in-depth interviews with members of the creative teams of 12 independent advertising agencies.

6.3.1. *Identification of the sample*

The reason for selecting agencies that are not part of any business conglomerate is due to the interest in studying the perception that independent professionals have of AI, assuming that, to the extent that companies belong to advertising groups or organizations, the use of AI tools would be conditioned by the shared criteria of their network of companies.

This choice comes with a significant limitation: the absence of a census that compiles all of the independent advertising agencies in Spain. Therefore, the study began with an initial effort to identify this type of company. To do this, the following steps were taken:

1) Exploratory design: the list of companies associated with the Club de Creatividad, Spain's leading advertising association, and its affiliated entities were consulted.

2) Sample configuration: a hundred companies were identified and contacted, inviting them to participate in semi-structured in-depth interviews, a type of interview known for its ability to obtain information through a professional conversation with one or more people for an analytical study [RUI 07].

3) Snowball strategy: the creative teams that participated in the interviews were asked for contacts from independent agencies that had not responded to the email, thus expanding the sample as much as possible.

4) Final database: finally, 12 creative teams from Spanish advertising agencies were interviewed and asked about their perception and use of GenAI in their daily work.

Interview 1 (E1)	Avante Medios
Interview 2 (E2)	Asomados
Interview 3 (I3)	BAP Conde
Interview 4 (E4)	Bululú
Interview 5 (E5)	Costa, Comunicación Visual
Interview 6 (E6)	Nasas
Interview 7 (E7)	Idea
Interview 8 (E8)	Imaxe
Interview 9 (E9)	Recados Carmen
Interview 10 (E10)	Reclam
Interview 11 (E11)	Roi Scroll
Interview 12 (E12)	Verve Creative Group

Table 6.1. *List of companies whose creative teams participated in the research*

The questionnaire, although semi-structured, had a series of fixed thematic blocks. Thus, the 12 interviews asked about 1) professionals' perceptions of GenAI; 2) the uses they give to these tools; 3) the importance advertisers attach to the use of AI; and 4) the limitations of these tools and their reservations about using them.

All participants had expert knowledge in their field [AUD 19] and freely expressed their opinions on GenAI, placing this work in the context of phenomenological social research, which gives scientific value to all perceived phenomena, whether objective or subjective [BAB 00].

In this specific case, the interviews were transcribed and then processed to organize and establish connections between the qualitative information collected [SAB 14]. To increase the quality and validity of the results, it was decided to eliminate the bias of a single researcher by using author triangulation. This technique consists of using a plurality of observers in contrast to the conventional methodology of a single observer [ROD 05]. In this research, three authors carried out the analysis of the transcripts. The reliability of the thematic analysis was verified by evaluating the level of agreement among the research team in the coding of the information. In this context, according to Cohen's Kappa coefficient (κ) [COH 60], the degree of agreement among the researchers is very high ($K = 0.91$).

6.4. Analysis of the results

6.4.1. *Perception of GenAI tools*

Most participants expressed enthusiasm for GenAI tools. "A revolution for the industry that has saved us a lot of work time and allows us to provide innovative production solutions" (E12).

Among those interviewed, the general trend is to highlight the operational efficiency of these applications, especially in content creation (text, images, videos, etc.). This is the fundamental contribution that participants appreciate in GenAI, although reference is also made to campaign optimization and the personalization of advertising experiences in real time. "In addition to creativity, there are many areas where Artificial Intelligence is bringing innovation to the advertising process: in identifying consumer niches, audience segmentation and campaign planning. It also provides objective data that allows you to plan a medium" (E1). "We are making advertising more personalised, adapting messages according to the behaviour, context or emotions of the target audience. This allows us to move towards more meaningful advertising, as well as improving engagement and return on investment" (E11).

In general, advertising agencies use GenAI programs on a daily basis, and many of them admit to developing their own tools that bring them business opportunities. "We have an engineer on staff. Initially, this person helped us with development activities, but over time they have specialised in helping us optimise processes. And not only internal agency processes, but also by developing innovative technological products that we offer to our clients. This makes us much more competitive as a company" (E5). This change in mindset is leading to the incorporation of new professional profiles into the advertising sector (mainly technicians and engineers) and the emergence of new business models (offering technological tools and communication solutions in addition to traditional advertising campaigns). "We combine communication profiles with technical profiles, and more and more clients are demanding digital communication actions supported by Artificial Intelligence solutions" (E2).

This enthusiasm coexists with a certain amount of skepticism and mistrust toward these advances, whether due to their fleeting implementation or their ethical implications. "Technology is progressing very quickly and every day a new application is born, so we still don't know where we are going to end up. I think GenAI is a very useful tool, but we should use it rationally in a context where society is misinformed and we all find it difficult to distinguish between what is true and what is not" (E3). Furthermore, we must not forget that there are hardly any regulations governing this practice, and professionals in the sector are calling for certain rules of the game to be established so that these solutions can be used in a consistent manner. "There are no clear criteria for regulating the use of GenAI. We try to follow good practices and be responsible communicators because we must be very aware that consumers need to be convinced through the truth" (E6).

Complementary to this are concerns about creative employment and debates about whether the use of technology will affect employability in the sector.

6.4.2. *Uses of GenAI in the professional environment*

The interviews revealed that GenAI has been fully integrated into the advertising production process, from the research phase to the creation of final artwork, including media planning. "Whenever we can, we try to include an GenAI I tool. In fact, we have an application to search for new business, so even before we get a client, we already attach great importance to this resource" (E9).

These solutions are present from the strategic conceptualization of a campaign, whether analyzing trends, identifying and profiling targets or conducting competitor analysis. "Market research is one of the most valuable intangibles our company offers, and we use AI to analyse large volumes of data, whether from social media, news, etc. We also use it to identify consumer behaviour patterns and analyse the competition or reconstruct their advertising campaigns" (E1).

During the idea generation phase, few creative teams claim to use this resource, but some say that AI is also useful for brainstorming. "ChatGPT is too tempting. I wouldn't think of using it to develop a campaign idea; that's something we already know how to do. However, it's very useful for imagining possible creative paths and thinking about different scenarios. It's another tool that helps us open our minds" (E12).

On the other hand, there is a certain unanimity in the use of these tools for image design and production. "GenAI is, above all, useful for sketching. Before, you would spend whole days looking through image banks for visual references or making a storyboard; now this has been greatly simplified because you can create the images you want in a very short period of time. What you present to the client is quite similar to what you will eventually produce" (E8). Thus, the generation of images and videos is the process in which this technology is most widely used, creating illustrations, animations or even complete videos from descriptions. "Also, for scriptwriting: once you have an idea, GenAI can help you write certain texts and identify the messages that should be added. Although you then have to work on them yourself because they tend to be somewhat clichéd and lacking in personality" (E4).

Sketching is one of the moments when this resource is most used, preparing models that help explain the idea to be realized. "We use GenAI to generate images of campaign layouts, but also for voiceovers. Before, you had to do the voice yourself or call a voiceover artist and bear the cost without knowing if you were going to do the campaign, but that no longer happens. GenAI generates quite good voiceovers, with a wide range of voices" (E9). But the content generated is not only developed at this stage. "We also produce with GenAI. In fact, we have a campaign that has received several awards and is almost entirely made with GenAI. We started with real human figures and had to go to a studio to record certain things, but most of the sets and the action that takes place are made with GenAI" (E8).

Finally, the use of this technology in the development and planning of campaigns is evident, whether it be automating repetitive processes, generating personalized experiences for each user, optimizing and correcting media planning, or managing insertions. "GenAI is very useful: we have incorporated it as another tool that helps us improve the performance of our campaigns, evaluating metrics such as CTR, conversion rates and engagement, identifying the ads that work best or predicting results that allow us to anticipate the impact a campaign will have" (E11). This confirms the incorporation of these resources into the entire advertising creation process.

As for the programs mentioned, ChatGPT is the most popular tool among participants, followed by MidJourney, Sora, Runway, ElevenLabs and Artlist. In the interviews, ChatGPT is referred to as the great icon of GenAI and an almost universal solution that is used for market research, generating ideas, writing texts and scripting. It is also used to create images, although in this case, MidJourney is widely favored among the creative teams interviewed, highlighting its ability to develop artistic concepts, illustrations and unique images, as well as its competence in developing consistent characters. For video production and post-production tasks, Sora and Runway were mentioned, while ElevenLabs and Artlist are used for soundtracks, transcriptions and audio solutions. Other names mentioned, albeit less frequently, were Canva, AdCreative, Copy, Pencil, Dall-E, Nano Banana and Adyouneed.

The main benefit highlighted by these tools is the saving of time and, consequently, money. "There is no doubt that GenAI simplifies processes, and tasks that used to take you many hours can now be completed in minutes" (E7). In addition, "saving time implicitly means saving money: because your teams can better manage their working hours and you no longer have to go to the sound studio if you don't know whether you are going to be awarded a campaign" (E12). Furthermore, some highlight the efficiency of these tools because "in certain repetitive design tasks, you reduce the margin of error, as GenAI is very useful for automating processes" (E5). Taken together, these are "aspects that facilitate our daily operations, but above all benefit the advertiser, whom we can serve in less time and at a lower cost" (E11).

6.4.3. *Advertisers and GenAI*

When assessing advertisers' attitudes toward GenAI, there is no clear consensus among the interviewees. The general impression is that advertisers welcome these tools with the same enthusiasm as agencies; "they use GenAI as much as we do because it makes their lives easier. It helps them summarise documents and save time. It also helps them develop strategies and automate processes. They even apply it to specific tasks such as active listening on social media or customer service" (E2). However, advertising professionals are wary of advertisers' expectations of GenAI.

The main criticism from professionals is its uncontrolled use.

> GenAI makes our lives easier. It saves us time, allows us to generate content that we could not have done otherwise, and enables us to create more personalised advertising that connects with what consumers expect. But in no way does it replace the human factor. We face the risk of people thinking that because they know how to use GenAI programmes, they no longer need an advertising agency (E12).

Along the same lines, there has been no shortage of opinions pointing to an impoverishment of the advertising process as a result of the indiscriminate use of these tools. "You go out on the street and you see more and more tacky adverts made with GenAI. I refuse to believe that this is what our clients want; we need professionals who can bring some judgement to a company's image" (E8).

Between these two perspectives, there are conciliatory views that predict a normalization in the use of these tools. "This is very new for everyone. We have only incorporated these technologies into our lives a few years ago and we are still trying out new routines. Advertisers and agencies are aware of their usefulness, but we need some time to find out in which areas they will help us and in which they will not. We are finding our place" (E10). These statements corroborate the rapid popularization of this technology in the advertising sector and the existence of numerous unknowns in its implementation process.

Some advertisers, for example, express certain reservations about incorporating GenAI into their campaigns because they like to do original things. "Advertisers like to see a sketch in GenAI. It helps them understand the idea, the audiovisual treatment and the production techniques proposed for a campaign. They can imagine it better, and this makes it easier for them to empathise with you. But then they don't want that to be their campaign: they ask you to do something unique and original" (E4). One point of view reinforces the contribution of GenAI in the strategy and sketching phases of campaigns but rules out its use in the final production of the pieces.

"Another interesting debate is about what is real and what is not" (E7). This is an issue that participants have given great importance to, because "there are many advertisers who do not like to show an artificial, imaginary image of their products. Nobody likes to lie to consumers" (E3). This is a growing concern in our society, which is highly sensitive to the manipulation of reality and misinformation, and which directly affects advertising, whose persuasive dimension has always led professionals to recreate aspirational worlds that are far removed from reality. Thus, the apocryphal identity of GenAI deters many advertisers from using it to generate images for their campaigns.

6.4.4. *Limitations and inhibitors*

The veracity of the generated image and its link to the existence of a true or apocryphal advertising discourse is a frequent topic when discussing the use of GenAI. However, it only acts as an inhibitor when launching an advertising campaign. "To a greater or lesser extent, we all use GenAI in our daily work, but not all clients dare or are able to run a campaign using AI. They don't mind if your sketches are done by ChatGPT, but they think twice before running a campaign with GenAI" (E4).

There is also concern about the impact of GenAI on creative employment. Many of those interviewed expressed concern about the possible loss of jobs in areas such as production and the realization of ideas. "There is no doubt that GenAI is becoming increasingly sophisticated, and it is unclear where we are headed" (E7). In general, participants do not believe that this technology will replace advertising professionals, "but we will have no choice but to adapt, because our work will change with the use of GenAI. There will be things that will no longer take up our time, and instead we will have to develop other skills" (E8).

Other ethical debates related to GenAI have to do with the role this technology is acquiring in our lives. "In the end, we are all a bit enslaved by GenAI. These tools should do the washing so that we can focus on doing our work, but sadly they do our work so that we can do the housework" (E12). In this dynamic, there are those who believe that we all lose out, as "there is a risk that with GenAI we will all end up being a bit the same, thinking in the same way and moving in common territories" (E3). This is where advertising professionals are needed, because "someone has to bring human intuition to the algorithm" (E7) to bring coherence to the process and allow advertising to remain an attractive and disruptive activity.

A final group of inhibitors that concern participants and weigh down their perception of these tools or discourage them from using them has to do with how difficult it is to keep up with new developments. "This is moving very fast, and you don't know if the programme you learned yesterday will be used tomorrow. None of us can be sure that the application we are working with will not be replaced in the coming months, and keeping up with technology is complicated" (E8).

6.5. Conclusions

Despite these reservations, there is no doubt that the advertising sector has incorporated GenAI as an important part of its professional routine. Advertising agencies, and particularly creative teams, make use of these tools, and the emergence of GenAI is a reality accepted by professionals.

This statement fulfils the main objective of the research, which is to analyze the degree of implementation of AI in the advertising creation process. The study of advertisers' perceptions of AI (O1) has allowed us to observe how participants welcome this new technology with a certain degree of enthusiasm, as it saves them time and money. The interviews reveal that AI is bringing about a change in the profession, demanding new professional skills, facilitating the incorporation of new profiles and changing business structures and models.

As for the main areas of activity in which GenAI (O2) is being developed, the campaign sketching process can be identified as the moment when its techniques have become most popular. Its contribution makes it possible to create campaign layouts or storyboards, as well as reference texts or voiceovers, in a simpler, more realistic way that is faithful to the creative idea. However, the interviewees do not limit its use to the sketching phase: they admit to using GenAI throughout the entire process, from the development of the campaign strategy to its implementation. These tools are also useful for identifying targets, preparing presentations, adapting campaigns and automating tasks, as well as for selecting media, optimizing real-time placements and predicting the impact of campaigns.

The new strategic conception of the profession implies a change in the mindset of the professionals interviewed, who admit to having had to modify their professional routines and acquire new skills. In this transition, inhibitors to the professional application of this type of tool have been identified, which reveal the concerns of advertisers and point to future avenues for improvement (O3).

The main psychological barriers to these applications have to do with the use of the real and the fake in advertising campaigns and doubts about how consumers will respond. There is also suspicion about the short-sightedness that some advertisers may suffer (believing that GenAI will make advertising agencies unnecessary) and how this may affect creative employment and the survival of the profession itself. These verbalized inhibitors show that professionals view this technology with as much uncertainty as enthusiasm.

Future studies could help to outline sectoral perceptions according to the type of company, considering any differences that may exist between full-service agencies, creative studios, digital agencies and media agencies. It should not be forgotten that in such a multidisciplinary sector, independent agencies specialize in different areas of advertising production and given the implementation of this technology throughout the entire process, it could be interesting to identify the different uses, perceptions and inhibitors verbalized by different professionals, such as advertisers who are at the forefront of the implementation of a new technology that is changing the way advertising was done until very recently.

6.6. References

[ATE 24] ATENCIO A., BARAHONA A., MONTERO I. et al., "Aplicación de la inteligencia artificial en los medios publicitarios", *Revista Semilla Científica*, vol. 6, pp. 659–668, 2024.

[AUD 19] AUDENHOVE L.V., DONDERS K., "Talking to people III: Expert interviews and elite interviews", in VAND DENBULCK H., PUPPIS M., DONDERS K. et al. (eds), *The Palgrave Handbook of Methods for Media Policy Research*, Palgrave Macmillan, Cham, pp. 179–197, 2019.

[BAB 00] BABBIE E., *Fundamentals of Social Research*, International Thomson Editors, Madrid, 2000.

[BAR 25] BARROLLETA L., FERNÁNDEZ E., ESPINOSA M., "Capítulo 2. inteligencia artificial en el periodismo, el marketing y la publicidad : una revisión sistématica de la literatura", *Espejo De Monografías De Comunicación Social*, vol. 36, pp. 33–53, 2025.

[BER 25] BERRÍO-MENESES C.M., SANGUINO-GARCÍA V., ISAZA-ÁLVAREZ J., "Artificial intelligence, data analytics and big data in marketing and customer and consumer segmentations. Systematic literature review", *Revista Galega De Economía*, vol. 34, no. 3, p. 10108, 2025.

[BLA 24] BLANCO SANGUINETI R.G., CARDENAS CORDOVA C.D., TORPOCO BALTAZAR A., "La inteligencia artificial en la publicidad : una revisión sistemática de la década 2020-2024", *Journal of the Academy*, vol. 11, pp. 53–82, 2024.

[CIL 24] CILLO P., RUBERA G., "Generative AI in innovation and marketing processes: A roadmap of research opportunities", *Journal of the Academy of Marketing Science*, vol. 53, no. 3, pp. 684–701, 2024.

[COH 60] COHEN J., "A coefficient of agreement for nominal scales", *Educational and Psychological Measurement*, vol. 20, no. 1, p. 37–46, 1960.

[DEL 24] DEL PINO-ROMERO C., ASENJO-MCCABE S., "El uso de la Inteligencia Artificial en el sector del marketing, la publicidad y los contenidos de marca en España", *Comunicación Revista Internacional de Comunicación Audiovisual Publicidad y Literatura*, vol. 2, no. 22, pp. 82–102, 2024.

[DEL 25] DEL PINO-ROMERO C., ASENJO-MCCABE S., HERRERA-DAMAS S., "El uso de la IA en la industria publicitaria en España : interés, conocimiento y uso", *Revista ICONO 14. Revista científica de Comunicación y Tecnologías emergentes*, vol. 23 no. 1, p. 2223, 2025.

[FER 25] FERNÁNDEZ CEREZO A., HIDALGO I., IZQUIERDO M., "La adopción de la inteligencia artificial en las empresas españolas : un primer análisis basado en la EBAE", *Boletín Económico, Banco de España*, vol. 06, pp. 1–12, 2025.

[FIE 24] FIEIRAS CEIDE C., VAZ ÁLVAREZ M., MAROTO GONZÁLEZ I., "AI implementation strategies in the Spanish Press Media: Organizational dynamics, application flows, uses and future trends", *Tripodos*, vol. 55, pp. 15–35, 2024.

[GRE 24] GREWAL D., SATORNINO C., DAVENPORT T. et al., "How generative AI is shaping the future of marketing", *Journal of the Academy of Marketing Science*, vol. 53, no. 3, pp. 702–722, 2024.

[HÄG 24] HÄGLUND E., BJÖRKLUND J., "AI-driven contextual advertising: Toward relevant messaging without personal data", *Journal of Current Issues & Research in Advertising*, vol. 45, no. 3, pp. 301–319, 2024.

[HER 25] HERNÁNDEZ-GÓMEZ O.S., FERNÁNDEZ-RINCÓN A.R., "Presente y futuro de la profesión publicitaria : la incorporación de la Inteligencia Artificial en el trabajo creative", in SÁNCHEZ-NAVARRO J., HELLÍN P. (eds), *Inteligencia Artificial : retos y oportunidades para la formación y el empleo en el ámbito de la comunicación*, Comunicación Social Ediciones y Publicaciones, Salamanca, pp. 287–310, 2025.

[INF 24] INFOADEX, *Estudio InfoAdex de la inversión publicitaria en España 2024*, InfoAdex, Madrid, 2024.

[JIM 25] JIMÉNEZ-JACINTO J.F., SEEBACH S., MARTÍNEZ-VALVERDE S., "La Inteligencia Artificial Generativa y la redefinición del proceso creative", in SÁNCHEZ-NAVARRO J., HELLÍN P. (eds), *Inteligencia Artificial : retos y oportunidades para la formación y el empleo en el ámbito de la comunicación*, Comunicación Social Ediciones y Publicaciones, Salamanca, pp. 147–170, 2025.

[KUM 24] KUMAR V., KOTLER P., GUPTA S. et al., "Generative AI in marketing: Promises, perils, and public policy implications", *Journal of Public Policy & Marketing*, vol. 44, no. 3, pp. 309–331, 2024.

[LAG 25] LAGUNA-GARCÍA M., NÚÑEZ-CASADO M.Á., MARTÍN-GARCÍA A. et al., "Capítulo 13. La IA desde la perspectiva de los profesionales de la publicidad : perfil de usuario y motivaciones", *Espejo De Monografías De Comunicación Social*, vol. 36, pp. 237–253, 2025.

[LAR 25] LA-ROSA L., ORTEGA-FERNÁNDEZ E., PERLADO M., "Inteligencia Artificial en el periodismo, el marketing y la publicidad : una revisión sistemática de la literature", in SÁNCHEZ-NAVARRO J., HELLÍN P. (eds), *Inteligencia Artificial: retos y oportunidades para la formación y el empleo en el ámbito de la comunicación*, Comunicación Social Ediciones y Publicaciones, Salamanca, pp. 31–58, 2025.

[LEI 25] LEITE F., "Inteligência artificial responsável e publicidade : pontos de atenção tecnoéticos para as demandas do presente e recodificação do futuro", *Intexto*, vol. 57, pp. 1–35, 2025.

[MAR 22] MARTÍNEZ MARTÍNEZ I.J., AGUADO J.M., SÁNCHEZ COBARRO P.D.H., "Smart Advertising : innovación y disrupción tecnológica asociadas a la IA en el ecosistema publicitario", *Revista Latina De Comunicación Social*, vol. 80, pp. 69–90, 2022.

[MAR 23] MARTÍNEZ-ROLÁN X., CORBACHO-VALENCIA J.M., PIÑEIRO-OTERO T., "Use of generative AIs in the digital communication and marketing sector in Spain", in MACHADO C.F., DAVIM J.P. (Coords.), *Management for Digital Transformation*, Springer, Cham, pp. 47–58, 2023.

[NOG 24] NOGUEIRA M.A.F., SOARES L.O., "A dicotomia entre criatividade humana e inteligência artificial no contexto da publicidade dirigida por dados : uma pesquisa exploratória", *Esferas*, vol. 14, no. 2, pp. 1–25, 2024.

[PER 25] PERUGACHI KINDLER A.E., "E-campañas : uso de big data, inteligencia artificial en campañas políticas híbridas y la aparición de "wikiliderazgos"", *Uru: Revista de Comunicación y Cultura*, vol. 11, pp. 63–85, 2025.

[ROD 05] RODRÍGUEZ RUIZ Ó., "La Triangulación como Estrategia de Investigación en Ciencias Sociales", *Revista madri+d. Monografía revista de investigación en gestión de la innovación y tecnología*, vol. 13, pp. 59–68, 2005.

[RUI 07] RUIZ OLABUÉNAGA J.I., *Qualitative Research Methodology*, University of Deusto, Bilbao, 2007.

[SAB 14] SABARIEGO PUIG M., VILÀ BAÑOS R., SANDÍN ESTEBAN M.P., "Qualitative data analysis with ATLAS.ti", *Reire: Journal of Innovation and Research in Education*, vol. 7, no. 2, pp. 119–133, 2014.

[SÁN 23] SÁNCHEZ-GARCÍA P., MERAYO-ÁLVAREZ N., CALVO-BARBERO C. et al., "Spanish technological development of artificial intelligence applied to journalism: Companies and tools for documentation, production and distribution of information", *Profesional De La información*, vol. 32, no. 2, pp. 1–18, 2023.

[SOL 25] SOLORZANO MARTÍNEZ D.P., OCHOA HERRERA V.A., YÁÑEZ CABRERA J.S., "Marketing y publicidad en la era de la inteligencia artificial y su impacto en la experiencia del consumidor", *Scripta Mundi*, vol. 4, no. 1, pp. 31–55, 2025.

[TOR 25] TORRES SÁEZ-DE-IBARRA A., MONTEAGUDO L., VÁZQUEZ T., "La integración de la IA en las agencias de publicidad, comunicación y en los medios de comunicación", in SÁNCHEZ-NAVARRO J., HELLÍN P. (eds), *Inteligencia Artificial : retos y oportunidades para la formación y el empleo en el ámbito de la comunicación*, Comunicación Social Ediciones y Publicaciones, Salamanca, pp. 311–334, 2025.

[TUR 25] TUR-VIÑES V., CASTELLÓ-MARTÍNEZ A., RAMOS-SERRANO M. et al., "AI, advertising, and the creative profession: Ethical challenges and professional transformations", in MIGUÉLEZ-JUAN B., REBOLLO-BUENO S. (eds), *AI Impacts on Branded Entertainment and Advertising*, IGI Global, London, pp. 89–108, 2025.

[UFA 23] UFARTE-RUIZ M.J., MURCIA-VERDÚ F.J., TÚÑEZ-LÓPEZ J.M., "Use of artificial intelligence in synthetic media: First newsrooms without journalists", *Profesional de la Información*, vol. 32, no. 2, pp. 1–17, 2023.

7

Emotional Nudging in the Rise of Affective Artificial Intelligence

This chapter examines how emotional nudging emerges at the intersection of affective artificial intelligence (AI) and behavioral choice architecture. It focuses on understanding how emotional nudging shapes human decision-making through affective intelligence and synthetic empathy. The analysis focuses on the mechanisms that enable emotional data to be sensed, generated and integrated into decision-making architectures, as well as their application across public policy, health, wealth and finance, education and sustainability. It also discusses ethical and societal implications and advances a set of guiding principles for responsible emotional nudging, aimed at safeguarding autonomy, privacy and fundamental rights.

7.1. Introduction: from nudging to AI-based emotional hypernudging

Human decision-making is inseparable from emotion. Far from being peripheral or contrary to rationality, as often seen in views centered on pure rationality, emotional processes are inherent in how individuals capture information, analyze it, weigh alternatives and form preferences. Contemporary research in behavioral economics and cognitive neuroscience has shown that emotional states influence valuation, risk perception and temporal discounting, thereby challenging the assumption of a purely deliberative *homo economicus* ([LOE 01], [LER 15]). Within this view, emotions operate not as noise but as integral components of reasoning, guiding attention and enabling adaptive choice in complex environments.

The concept of *nudging* emerges from this recognition of limited rationality, combining the insights of behavioral economics with its potential applications in shaping public policy. Defined as a subtle modification of the choice architecture

---

Chapter written by Cristiana Cerqueira LEAL and Benilde OLIVEIRA.

that alters behavior predictably without restricting freedom of choice [THA 08], nudging relies on cognitive and affective heuristics that drive behavior in real contexts.

Nudges have gained widespread recognition for their application in public policy to improve individuals' choices in their own best interests as if made by them in a thoughtful and deliberate manner, without forbidding or altering economic incentives, thereby contributing to individual well-being and, by extension, that of society. The goal of nudging is to use choice architecture for good, and it has been widely accepted around the world by governments, government agencies and non-governmental organizations (see, for instance, [WHI 14], [LEA 21]).

Traditional nudging interventions, typically developed in behavioral economics and public policy, are often designed for a specific population and implemented in a homogeneous way for all [LEA 24]. Such uniform approaches can be effective for certain groups but may fail to account for the heterogeneity of individual preferences, cognitive styles or socio-emotional contexts. Consequently, the behavioral impact of traditional nudges is bounded by the assumption that a single intervention can produce similar outcomes across diverse populations.

With the digitization of decision environments, *digital nudging* emerged as the application of the same behavioral principles within interactive technologies [WEI 16]. Digital tools enable users to be tracked through various means, including cookies, profile settings in applications, account registrations on multiple platforms, and the integration of this information in a coordinated manner to create individual profiles. Then, digital nudges can be personalized for individual characteristics in a dynamic and adaptive way, by exploiting the data-rich, iterative nature of online interactions [LEA 25]. Personalization becomes possible, allowing nudges to align with individual traits, decision histories and contextual variables.

With the widespread use of digital tools based on AI, a new paradigm has emerged in big data processing, user profiling and feature customization. Each user engagement provides new feedback that can inform algorithmic learning and incremental refinement of the intervention. Personalization becomes possible, allowing nudges to align with individual traits, decision histories and contextual variables. Digital nudging thus transforms the nudge from a static policy instrument into an evolving, data-driven mechanism embedded in the architecture of digital interfaces that update and refine continuously in real time: the hypernudging.

Building upon adaptive AI-based capability, *hypernudging* represents a further step in complexity and autonomy. Coined by [YEU 17], the term denotes systems in which nudging algorithms operate simultaneously, continuously adapting the digital choice environment in real time. These systems use big data analytics, machine learning and automated feedback loops to refine influence strategies across a wide range of behavioral domains, including financial decision-making, consumption and information exposure. Hypernudging is thus characterized by scalability, algorithmic responsiveness and the capacity for continuous self-optimization, creating a fluid architecture of influence that evolves with each interaction [LEA 24].

In the context of AI, emotional nudging extends these innovations into the most recent and expected to expand exponentially in the near future, affective dimension of human–AI interaction. *Emotional nudging* is a way of gently appealing to people's emotions to guide how they think, decide and act. From all behavioral insights, emotional nudging focuses on using emotional cues and affective aspects in the choice architecture to produce behavioral change in a subtle and predictable manner. It intentionally uses emotional cues (linguistic, visual, auditory or behavioral) to influence decision processes rather than relying on purely cognitive heuristics. Then, emotional nudging is a strategic approach that deliberately targets individuals' emotions to influence their decisions and actions. Rather than focusing on improving rational decision-making, this approach recognizes that emotions play a primary role in shaping individual behaviors and uses that to conduct decision-making in a premeditated direction.

The rise of *affective computing and emotional artificial intelligence (AI)* has made such interventions increasingly feasible. Affective computing, as coined by [PIC 00], aims to equip computational systems with the capacity to detect, interpret and express emotions. Through advances in multimodal emotion recognition, voice and facial analysis, and natural language processing (NLP), AI systems are now capable of dynamically adapting interactions to users' affective states ([CAL 10], [MCS 18]). This technological progression enables what may be described as *algorithmic empathy* or *synthetic empathy*, referring to the emerging simulation of empathic responsiveness by machines [HO 23].

Emotional nudging, therefore, represents the convergence of behavioral science and affect-aware computation: it operationalizes emotional intelligence in algorithmic form, enabling choice architectures that adapt not only to cognitive patterns but also to *dynamic emotional feedback*. It appeals to emotionally driven events that resonate on a personal level and prioritizes the feelings and experiences associated with specific contexts, such as health or wealth decisions.

An emotional nudge provided by AI systems implies approaches that read, categorize and react to human emotions. They rely on data about an individual's emotional state, which may include biometric data such as pulse, heart rate, eye, head and body movements, as well as facial expressions. For instance, in video conferences, participants can be analyzed in real time based on eye and head movements, as well as microfacial expressions, to determine their emotional state, which can be classified in terms of attention, understanding and agreement throughout the conference. AI systems may continuously evaluate the user's affective state and update an intervention policy that times and frames particular elements (ordering, salience, defaults, visual aids, speech rate, tone of voice) to steer engagement. AI emotional sensitivity within digital environments can adjust tone, framing or presentation in response to emotional cues. Such affective adaptation opens new pathways for influencing decision processes in areas ranging from financial behavior to health communication.

Obviously, this immense progress in the way human beings are observed, interpreted and shaped through emotions does not come without ethical and legal challenges. This knowledge can be used for obvious improvements in decision-making processes, but it comes with inevitable losses of privacy, potential losses of autonomy and a risk of harmful manipulation.

This chapter presents an integrated perspective that combines three strands of research often treated separately: behavioral economics and nudging, emotional and affective AI, and the emerging debates on AI ethics and regulation. Building on this synthesis, it (1) defines emotional nudging as the convergence of affective AI with behavioral choice architecture; (2) proposes an operational three-block model – recognition, expression and adaptive personalization – complemented by temporal and contextual variability; (3) maps cross-sector applications in public policy, health, sustainability, finance, education and digital platforms; and (4) advances a set of guiding principles for responsible emotional nudging that integrate AI-ethics frameworks and current regulatory developments with behavioral insights.

7.2. Emotions and decision-making

7.2.1. *Human emotions: what is an emotion?*

Emotions are a defining feature of human experience and a primary factor shaping how the brain processes and responds to the surrounding world. They influence perception, attention, memory and judgment, guiding behavior through mechanisms that are rapid, automatic and evolutionarily adaptive [LED 12]. Far from being peripheral to cognition, emotions operate as a biological and cognitive interface between the body and the environment, enabling the organism to prioritize relevant information and prepare for action.

Understanding emotions first requires a clear definition of what constitutes an emotion. Although the term is a familiar concept in everyday language, philosophers and scientists have attempted to define, categorize and map emotional phenomena. From Darwin's *The Expression of the Emotions in Man and Animals* [DAR 72] to contemporary affective neuroscience, the effort to identify the essence and function of emotion has not yet reached a definitive consensus.

From a behavioral insight standpoint, what matters most is that emotions are primary drivers of human action. As António Damásio argues in *Looking for Spinoza* [DAM 03], *emotions precede feelings*; they are the body's first and most immediate response to internal or external stimuli. Feelings, in turn, are the mental representation of those emotions in the moment when the individual becomes conscious of their emotional state. In this sense, emotions are not the consequence of thought but its precursor: they orient cognition by signaling what matters to the individual before deliberate reasoning even begins [DAM 18].

For choice architecture design, this insight is crucial. Decision-making is rarely the result of rational analysis alone; rather, it is an emotionally charged process in which affective cues shape preferences, expectations and perceptions of risk. Emotions influence how information is framed and evaluated, as well as how alternatives are weighed under uncertainty [LOE 01]. Fear, for instance, can lead to loss aversion; joy can encourage risk-taking; regret can alter future choices. The emotional context acts as a lens through which the rational mind interprets value and probability. Emotions thus perform a regulatory function in human behavior. They simplify the complexity of decision environments by attaching an affective charge – positive or negative – to experiences and anticipated outcomes. This capacity to assign emotional value gives direction to human choice, grounding it in motivation rather than calculation.

Therefore, emotions are a primary and constant element that can trigger, accompany, instigate and reinforce rational thinking. They constitute an effective underlying force in shaping any course of action. As [DAM 18] observed, *emotion and reason* are not opposing forces but complementary systems within the same architecture of the mind. Reason, stripped of emotion, becomes sterile; emotion, without reason, becomes impulsive. It is in the dynamic balance between both that human judgment takes form. In this light, emotions should not be regarded as distortions of rationality but as its foundation. They are the first evaluative layer through which individuals navigate the uncertainty of real life – the quiet yet persistent compass that orients what feels right, not only what seems logical.

7.2.2. A catalogue of emotions for decision-making

Categorizing emotions can be even more challenging, and despite several attempts, the task remains complex and incomplete. The most useful way to "catalogue" emotions is to start with the small cluster, the so-called *basic or primary emotions*[1], that show the greatest cross-cultural regularity and the clearest links to action: fear, anger, joy, sadness, disgust and surprise [EKM 92]. These are typically fast, automatic and tightly coupled to approach–avoidance tendencies and attentional shifts – precisely the levers that shape choices under time, uncertainty and limited cognition:

– *Fear* prioritizes threat detection and caution. In economic contexts, it tilts perception toward losses and downside risk, often amplifying avoidance and insurance-type behaviors.

– *Anger* mobilizes approach and norm enforcement. Unlike fear, anger can reduce perceived risk and increase willingness to act, sometimes fostering risk-taking when blame is salient.

– *Joy or happiness* broadens attention and promotes exploration. It is associated with optimism, higher willingness-to-pay for experiential goods and a greater tolerance for variance in outcomes.

– *Sadness* narrows the field of view to immediate relief and restitution, with downstream effects on intertemporal trade-offs and price sensitivity.

– *Disgust* produces sharp avoidance and devaluation, spreading quickly by association (important in health, food and ethical consumption).

– *Surprise* functions as a rapid attentional reset that updates priors and can open a transient window for reframing and preference revision.

These are emphasized because they provide stable anchors for predicting shifts in evaluation, attention and preference, which are relevant for modeling behavior in markets and everyday economic decisions.

Beyond this core, the catalogue becomes more *emotion-graded and context-dependent*. Emotions such as guilt, shame, pride, envy, gratitude, compassion, awe, hope, regret, embarrassment and contempt often involve richer appraisals of agency, norms and social evaluation [COW 17]. They matter greatly for domains such as tax compliance, prosocial giving, sustainable consumption and long-term saving. Nevertheless, their boundaries are less precise and their measurement more sensitive to culture and situation. In practice, these emotions can add explanatory power when *social evaluation* or *moral salience* is central to the decision environment.

1 Depending on the source, the basic list of emotions can range from 5 (in which surprise is the one not included in this list) to 7, 8 or more, adding emotions as anticipation, trust or contempt. Ekman, in his most recent research, tends to recognize contempt as a seventh basic emotion.

The quest for understanding and exhaustive cataloging of emotions is not a recent endeavor in the literature. In the 17th century, *Spinoza's Ethics* (Part III, *De Affectibus*) treated emotions as primary drivers of action rather than mere surface manifestations. Starting from *conatus* (from Latin *cōnātus* meaning attempt, endeavor, effort, i.e. the inherent striving to persist in existence), he distinguishes emotions from feelings and moods and derives a *systematic catalogue of 48 affects* in the *Definitions of the Emotions,* by combining shifts in the power to act (joy/sadness) with ideas about their causes [SPI 02]. Read through today's lens, this framework anticipates the operational logic of emotional nudging: identify the state, anticipate the action tendency and shape the choice context accordingly. In effect, Spinoza provides an early template for linking affect to attention, valuation and behavior. [DAM 03] revisited this linkage, arguing that emotions precede feelings and provide the primary regulatory layer for thought and behavior. Contemporary AI-based affection models now implement this framework and metrics in adaptive interfaces for behavioral change.

Hence, three complementary lenses help organize affect catalogues for emotional nudging design:

– *Basic-emotion accounts* offer parsimony: a small set of states with directional predictions. For example, fear implies greater risk aversion, or anger implies lower perceived risk and higher readiness to act [EKM 92]. These mappings support targeted nudges that time and frame choices to align with the expected action–reaction tendency.

– *Dimensional models*: treat affect as emotional tone and intensity (seen as coordinates in a valence–arousal space), rather than labels, explain variation in choices. In a dimensional nudge, set a target in the valence–arousal space (e.g. moderately positive, medium arousal), score user message wording for tone intensity with *natural language processing (NLP)*, deploy the closest variants and adapt in real time based on behavioral lift against target metrics.

– *High-resolution semantic maps*: reveal families of related emotions bridged by continuous gradients. They support finer distinctions, such as joy versus awe, when testing mechanisms in specific contexts such as charitable appeals or experiential purchases [COW 17].

The operative rule is to *match user granularity to the nudge objective*. For scalable prediction and clean comparability in field experiments, the *basic set* usually suffices. When the mechanism activates norms, identity or reputation, extend the palette to include *self-conscious and social emotions*, with measurement validated for the target population and context. When *intensity and tone* are the primary levers (e.g. sustaining engagement or softening resistance), dimensional metrics capture the necessary variance with fewer assumptions, allowing for precise and real-time adjustment of emotional framing.

7.3. Mechanisms of emotional nudging through AI

7.3.1. *Human emotions versus synthetic emotion in AI*

We will not focus the discussion on whether AI, as a machine, can experience emotions, in the sense of reproducing the neurological processes of emotions and faithfully experiencing emotions. We believe that in the near future, with the evolution of knowledge about emotional processes as neuronal processes, as well as the possibility of capturing and integrating more information from the surrounding environment, emotions simulated by AI machines will tend to become increasingly similar to innate human emotions. In this sense, it may become increasingly challenging to distinguish simulated emotions from real emotions, which is different from saying that the machine is genuinely (i.e. physically) experiencing the emotion.

The purpose of emotional nudging through AI is not to make the system feel as human, where the discussion about AI having the ability to experience real emotions would be central. On the contrary, in emotional nudging, the goal is for the AI mechanism to have the ability to detect, interpret and respond to human *emotions in a synthetic empathetic way*, thereby reshaping how decisions are made across various contexts. While affective computing traditionally describes these steps functionally, here we explicitly reinterpret them as choice-architecture levers, turning emotional AI pipelines into a nudge design space.

Emotional nudging, when implemented in intelligent systems, follows a simple logic: sense the emotional state, shape the stimulus accordingly and learn from the outcome to fine-tune future interactions. The objective is not to replace deliberation but to stage the choice context so that attention, valuation and action tendencies move in predictable directions with minimal friction. What matters most is to read the person's emotional state and respond with empathy that feels authentic: the right tone, at the right intensity, to invite acceptance rather than trigger resistance by signaling manipulation.

In short, the practical question in nudging domains is not whether a machine "feels" but whether it can functionally align with human affect in ways that ease judgment (rather than distort it). What counts in emotional nudging is calibration: detecting tone and intensity with enough fidelity, responding with language and pacing that feel natural, and maintaining that congruence across interactions so that the exchange reads as sincere rather than contrived. When this alignment holds, emotions become a workable input to choice: reducing friction, clarifying priorities and guiding action in the user's best interest, without narrowing the option set. Together, these elements turn emotional information into a controllable input for choice architecture that is measurable, comparable and ready for cumulative improvement. Then, three operative blocks are needed for modeling behavior: recognition, expression and adaptive personalization.

7.3.2. Emotion recognition: reading the room

The identification of emotions follows the ideas initiated by Spinoza in the 17th century [SPI 02], which he called *treating "human vice and folly geometrically"*, referring to his framework of postulates, propositions and definitions of the emotions. This framework is now using more comprehensive and in-depth strategies supported by AI.

For nudging to be timed and toned effectively, the system needs a workable estimate of the *user's emotional state*. In practice, this means translating observable cues, such as language, voice, facial expressions or simple interaction traces, into a compact affect signal that policy can use. The aim of emotional recognition is not clinical diagnosis but an actionable signal (such as "negative tone, moderate arousal" or "rising frustration") sufficient to adjust framing the wording, slowing the pacing or adding a summary and recap [CAL 10], [PAN 23], [GEO 24].

Multimodal approaches improve reliability by combining channels. A banking app can combine text sentiment with a click flow friction (time to complete, retries) to detect anxiety during a transfer; a support bot can combine prosody (rising tone, short responses) with content (increasingly negative phrases) to detect anger. A learning platform can track long pauses and repeated hints as a sign of confusion. When camera access is available and appropriate, transient facial cues, such as raised eyebrows in surprise or pressed lips in frustration, can add a slight additional layer of confirmation.

Emotion can be assessed with a toolkit that adapts to the context and can be based on several approaches, namely:

– *Self-reported emotional state* uses brief items, such as a mood slider with a question or short valence–arousal ratings placed at key moments, embedded at key moments to anchor other signals.

– *Traces of attention and disengagement* read in engagement from simple interaction data such as search terms, clicks, scroll depth, dwell time and inactivity, session time, device or application used, hesitation latency, and forward and backward patterns. These features represent focus of attention, effort and avoidance.

– *Linguistic analysis (text signals)* that assesses tone directly from the text, extracting sentiment and emotion, as well as pragmatic cues, such as openness, certainty, politeness and frustration. Linguistic analysis uses natural language processing (NLP), including large language models (LLMs), where appropriate, to infer and score these features at scale. This approach combines lexicon-based and contextual embeddings with domain fine-tuning, human-rated calibration and periodic validation to mitigate bias and overconfidence in real-world applications.

– *Observation of behavioral cues* through various mechanisms such as fleeting facial expressions or looking at the camera, vocal characteristics (tone, intensity, speed), posture changes and simple interaction traits such as hesitations, retries, back clicks or time to complete.

– *Physiological measures*, when appropriate, may capture readings as heart rate variability for regulation/stress and pupil dilation (i.e. pupillometry, a sympathetic arousal index measurable via eye-tracking) or electroencephalography (EEG), which is reserved for time-constrained lab tasks.

– *Multimodal emotion recognition* combines several strategies (e.g. text + prosody + click flow) and triangulates these channels to increase reliability before adjusting tone, pace or framing.

Since estimates are uncertain and context-dependent, robust implementations add quick checks for users' satisfaction (for example, for an item's mood control, a message such as "*Was this helpful?*" or a brief re-engagement task). Typically, implementations use conservative thresholds before adjusting their strategy. The recognition block then provides a state variable (a succinct, continuously updated signal) that downstream AI components use to decide when to intervene and how to embellish the message (e.g. a warmer tone, slower pace or a confidence-inspiring summary), keeping the intervention proportional to the moment.

7.3.3. *Emotion expression/generation: shaping the stimulus*

Once a target state is set (e.g. moderately positive and medium arousal), the AI system shifts from reading emotions to *shaping how the message lands.* In text or voice, this means adjusting wording and prosody to fit the moment. For instance, a vaccination reminder can move from "*Book your appointment*" to "*You're almost protected. Book today and keep your family safe*" with the voice slowing slightly, softening emphasis and closing on a calm, rising contour; a tax notice sent during peak stress can lower tempo and intensity, acknowledge strain with the emphatic message "*Let's take this step by step*" and present the same options more gently; if attention wanes in a recycling or energy-saving campaign, a short positive recap such as "*One small change, one big impact this week...*" helps restore focus without adding friction [CAL 10].

Visual and embedded interfaces offer parallel levers for policy implementation. A tax portal can reduce clutter, warm up the color palette and add white space to ease emotional load, while a public health app can combine a reassuring chatbot with a single salient call to action, such as "*Check eligibility*".

A practical rule is to *keep the content constant while varying the emotional envelope*. This is the core of emotional nudging: modulating salience, timing and affective framing, rather than incentives or the set of choices ([PIC 00], [WEI 16], [JES 21], [GAS 24]). In practice, it all comes down to small, testable adjustments (as tone, pace, emphasis and layout) that increase receptivity and compliance with behaviors such as vaccination, timely filing, safer travel or reduced energy consumption, while leaving the decision itself intact.

7.3.4. *Emotional personalization in adaptive loops: what works for whom*

Human emotions are heterogeneous. Accordingly, emotional responses are heterogeneous too. Affective AI systems, therefore, maintain lightweight profiles of users' responsivity (e.g. sensitivity to loss-framed vs. gain-framed messages, tolerance for arousal) and update them with each interaction and a personalization loop (for further detail on nudging personalization features, see, for instance, [LEA 24]). In practice, this means testing small variants and letting evidence guide exposure. A vaccination campaign might compare a gain frame like "*Protect your community. Book today*" with a loss frame like "*Delaying leaves you at risk*", then route more reminders to the variant that lifts bookings for a given neighborhood or age group. A tax portal can trial calmer versus brisker prosody in voice assistance during peak season and gradually favor the tone that reduces abandonment. Energy programs can alternate between goal-focused and social-norm messages, such as "Households like yours saved 8% last month" and adjust delivery to favor the option that reduces usage.

The feedback loop is simple and auditable: (1) estimate state; (2) deliver a variant; (3) observe behavior (booking, filing completion, dwell time, opt-in); and (4) update the policy to reflect the individual emotional profiles and keep, continuously, the loop of tracking and personally adjusting the emotional nudge to conform with human emotional resonance in the human–AI interaction.

This marks the era of affective computing and empathetic AI personal assistants, where adaptive mechanisms tailor hypernudges to users' emotional profiles. Using emotional profiling grounded in prior affective responses, systems continuously sense emotions, predict user reactions and adjust presentation via real-time reinforcement-learning loops that optimize both timing and tone. The goal of optimization is not "emotion" in the abstract, but rather individual behavior conditioned by emotional state. Emotion is merely a means of personalization to reach the target more quickly and efficiently, and to maintain that efficiency over time.

7.3.5. *Temporal and contextual variability*

Emotions are not static labels but moving targets: they rise, peak and decay on different timescales, and their impact on choice depends on where the person is in that trajectory and what situation frames the episode. Affective science describes this as variation in time course (onset, intensity, recovery) and contextual embedding (who, where, what is at stake). In practical terms, fear that spikes during a safety alert is short-lived and action-oriented; low-grade anxiety during a tax deadline is slower, more pervasive and tends to tilt evaluation toward caution. The same nominal "emotion" therefore yields different behavioral signatures across time and setting ([KEL 10], [LER 15]).

In emotional nudging, hyper-personalization channeling into a narrow range of cues may increase efficiency in the short term. However, repeatedly serving the best-performing variant for a given state-context pair may eventually lead to fatigue and inefficiency in the long run. The nudging intervention risks overfitting to a specific optimum and becoming less responsive to context shifts, consequently restricting the experiential range of cues offered to the user, with predictable implications for autonomy.

Adaptive hypernudges amplify this dynamic: by rapidly testing and locking onto the top performer, they can improve short-run efficiency yet funnel users into a narrow band of cues, overfitting to a specific optimum and becoming inelastic over time and in the face of context shifts.

The context constrains the same categorization differently in different areas of emotional nudging use. For instance, a state of moderate and slightly negative arousal may invite calmness in health messages, clarity under time pressure in transportation or loss mitigation framing in tax reminders. Attaching compact context keys (task, risks, channel, seasonality) to each observation allows policies to remain state-conditional and context-specific.

Framed this way, effective emotional nudging has less to do with defining a category and more to do with timing and appropriateness of context, aligning minor adjustments in tone, pace and salience with the direction of the state and what the situation demands.

7.4. Applications of emotional nudging

7.4.1. *Public policy and civic engagement*

Emotional nudging supports compliance and prosocial application when policies face attention limits or mild resistance. Practical targets include on-time tax filing, booking vaccinations, saving energy, participating in recycling and preparing for emergencies.

This means, for instance, calming tax and benefit communications during peak stress so that citizens can complete forms rather than abandon them; framing vaccination reminders around reassurance and collective care to increase bookings; and pairing energy-alert messages with a sense of local contribution to nudge short-term conservation. Campaigns work best when cues are woven into default touchpoints, such as portals or SMS reminders, and timed to moments of hesitation or overload, so the message feels like help rather than pressure.

As behaviors are often formed from habit and tend to be repeated in similar contexts, small, well-timed adjustments in default public portals or reminders can shift outcomes [LEA 22]. Public choice architecture initiatives demonstrate that when these small adjustments are incorporated as facilitators into existing environments, rather than implemented as one-off measures, they yield more stable outcomes and adherence over time. For example, policy designs that match emotions to specific moments: reassurance and efficacy for health appointments, loss-mitigating frames for deadline communications, collective pride and local identity for civic drives, and calm, high-clarity messaging during disruptions. Social-norm field studies in utilities reveal a similar pattern: lightly affective comparisons, such as "*households like yours reduced usage last week*", yield measurable, durable savings without resorting to heavy-handed appeals [ALL 11].

7.4.2. *Health and well-being*

Emotional nudging helps public-health messages land at the right moment and in the right tone – supportive, efficacy-focused and non-moralizing – so people move from intention to action in promoting healthy lifestyles. In preventive health and chronic-care routines, emotional nudges help sustain adherence (screenings, medication, physical activity, sleep hygiene) and reduce relapse. Reminders that acknowledge effort and offer a clear next step, lift bookings and cut no-shows (e.g. "*You've done the hard part! Book your slot; it takes two minutes.*"). During lapses, recovery messages bring people back without shame (e.g. "*Missed a dose? Pick up tonight; your plan stays on track.*"). For lifestyle change, brief celebrations of progress help maintain momentum (e.g. "*Four days with 1+hours of activity! Keep the streak.*").

Large-scale field studies demonstrate that simple, well-timed messages – framed around ease, social support or personal commitment – can increase the uptake of vaccinations and other preventive behaviors without heavy-handed pressure [MIL 21]. In ongoing routines, combining small goals with supportive check-ins improves persistence in physical activity, especially when individuals receive timely feedback on their progress [PAT 16]. For emotion nudges design, prior syntheses emphasize embedding these cues in default touchpoints (e.g. public portals, primary care outreach, pharmacy apps) and maintaining a tone that is helpful rather than urgent, which improves acceptability and adherence to healthier lifestyles in the long run.

7.4.3. *Sustainability and climate action*

Emotional nudging helps turn broad sustainability and climate intent into everyday choices that stick. In a domestic setting, it can frame consumption and renovation decisions (such as insulation, heat pumps and glazing) that prioritize comfort, health and satisfaction in a more welcoming and sustainable space. In retail, messages that encourage repair first and positive suggestions about refilling and reusing make the eco-friendlier and waste-free option seem rewarding. In terms of mobility, route planning and carpooling, as well as the use of public or soft mobility, highlight smooth and predictable journeys and the sense of contributing to cleaner air in the neighborhood and strengthening community life. Municipal programs can also recognize community milestones, such as the use of public transportation or soft mobility, using pride and gratitude to sustain the effort over time. Policy toolkits are most effective when these cues align with existing instruments, including clear disclosure, sensible defaults and simple incentives [PIC 08], [LEA 21]. The guiding principle is to keep options open while using emotion to reduce friction and maintain momentum. In parallel, financial channels can support sustainable choices, from green mortgages for energy-efficient homes to ESG (Environmental, Social, and Governance)-oriented investing, fostering individuals to make a commitment to sustainability in their spending, saving and investing that incorporates these choices on a lasting emotional level [OLI 24].

7.4.4. *Financial behavior*

Emotional nudging is a natural fit for everyday money choices, where inertia, anxiety and short-term focus often derail good intentions. In practice, this is evident in three dimensions: 1) *onboarding and contribution*: gentle, efficacy-focused prompts at predictable moments (e.g. just after payday) help people initiate or raise automatic transfers without pressure; brief, confidence-building explanations reduce hesitation around diversified funds or pension defaults; 2) *staying the course*: recognizing milestones supports persistence. During market swings, calming language and concise rationales (e.g. "*Here is why this mix still fits your horizon.*") curb knee-jerk selling and sustain contribution plans; 3) *maintaining resilience*: low-arousal, high-clarity alerts reduce scam losses and bill-payment errors, while short check-ins keep budgets on track. The policy value comes from pairing these cues with proven instruments, such as auto-enrolment, smart defaults and simple commitments, so that emotion lowers friction rather than replaces structure [MAD 01], [THA 04].

AI-powered *personal assistants and recommender systems* extend this approach at scale. AI-powered financial assistants can process a vast pool of personal financial data and sequence individuals' financial microtasks to create highly tailored nudging interventions by designing new features for convenience and interaction. These personal financial assistants can be leveraged by *social (ro)bots* that can pace decisions with calm, efficacy-focused prompts at paycheck cycles and provide brief, positively toned recaps when users hesitate on transfers or bill payments. In savings and budgeting, AI financial assistants that acknowledge stress can offer one-click next actions and achieve higher follow-through rates than neutral reminders. AI-based financial recommender systems extend this logic to investment portfolios and financial products: explainers framed in confidence-building, non-urgent language increase adoption of default contributions and diversified funds, while loss-heavy or urgent tones depress uptake among risk-sensitive users [LEA 25], [OLI 25].

7.4.5. *Education and learning environments*

In education, emotional nudging supports engagement (returning to the platform, time-on-task), persistence (reducing abandonment after errors) and timely submissions. In classrooms and online platforms, brief, encouraging messages that normalize difficulties help maintain effort on problem sets and projects. Rather than imposing more difficult content or tighter deadlines, the emphasis is placed on *tone, pace* and *timely recognition* of effort, so that students return to the task rather than disengaging.

The design of educational programs benefits from insights gained from the literature on emotion: emotions such as confusion and frustration can lead to a tendency to procrastinate or give up, while clarity and satisfaction are associated with increased attention and persistence. Success can be promoted by the way tasks are designed and feedback is framed. Learning path designs that mix achievable steps with a calm tone tend to prolong the time spent on the task and improve the quality of results. These supportive cues should be implemented especially at known dropout points in the learning process [PEK 06].

AI-based learning platforms and *AI tutors* can operationalize emotional prompts by modeling each learner's profile, using signals such as error patterns, latency, hint use and reattempts, and, accordingly, triggering personalized cues at the right moment and intensity to guide the learning process. These small, timely stimuli that acknowledge difficulties and celebrate incremental progress are associated with better learning trajectories, as they reduce the emotional friction that causes learners to abandon a learning sequence [DME 12]. For institutions, the practical markers of success are straightforward: fewer mid-module drop-offs, steadier participation and higher task completion rates. The through-line is simple: use light, well-placed emotional cues to make continuation feel doable, keeping the choice set intact while nudging students toward the next learning step.

7.4.6. *Digital platforms: social media, e-commerce, fintech*

On large *digital platforms*, emotional nudging is executed by AI systems that observe high-resolution behavioral data, such as search terms, scroll and pause patterns, dwell time, purchase funnels, chat tone, and even time-of-day rhythms, and translate them into granular, person-specific experiences. This scale enables hypernudging: many small, concurrent adjustments (ranking, defaults, copy, color, cadence, prosody) are tested and re-weighted in near real-time for each user segment or individual. The result is an interaction that feels tailored, as it is continuously re-estimated based on recent signals and context [YEU 17], [JES 21].

On *social media*, affective systems modulate what is displayed and how it is framed. Recommendation and response assistance models learn which emotional tones drive user engagement on a given topic and subtly exploit them, validating and reinforcing what generates engagement. Social bots amplify this dynamic by applying emotionally charged content with individualized and interactive personalization, altering perceived norms and attention flows in ways that change users' choices [FER 16].

In *e-commerce*, similar AI systems operate on sales platforms that connect with users by triggering emotions to build brand recognition and induce purchasing decisions, without changing the set of options. In *fintech*, emotionally appropriate explanations and social bots affect adherence to savings plans, investment portfolio management and fraud prevention. Virtual financial assistants and recommender systems are particularly effective when they combine language focused on effectiveness with clear, simple steps to follow and a long-term goal orientation [LEA 24].

Large digital platforms act as *gatekeepers*, with access to enormous amounts of user data that allow them to control how emotional stimuli are designed and delivered (for further discussion on these gatekeepers, see [LEA 25]). Operating systems, app stores, identity, cloud, ad exchanges and core recommender infrastructure are at their disposal to cross-reference information that enables a series of cross-hypernudges across multiple dimensions and the integration of user life across multiple platforms. Control at these layers means control over what is measurable, where interventions are placed, and how quickly feedback loops learn. Emotional influence is the most delicate layer of that control: tuning tone, salience and pacing can redirect attention and shape preferences while keeping choices intact on paper. At platform scale, this produces powerful distribution effects – whose strength comes not from any single message but from the accumulation of many small, affect-aware adjustments across feeds, search, checkout and assistance.

7.4.7. *Personal assistants*

AI personal assistants, whether in the form of voice or chat on an app device or embedded in services, are a natural environment for emotional nudging because they sit at decision moments, such as planning the day, paying bills, choosing transportation, or sticking to health or budgeting goals. They can intersect with various aspects of the user's life, including work, leisure, finances, shopping and traveling. Assistants can provide timely tips, facilitate interaction and transform intentions into actions, based on the continuous inputs they receive to assist in each specific situation. What differentiates assistants from static interfaces is the relational layer: they sustain engagement by acknowledging effort, matching tone and pacing to the moment, and offering a clear next step aligned with a long-term vision, step by step. Evidence on *relational agents* shows that a calm, empathic stance maintains user commitment over time more reliably than neutral, transactional dialogue [BIC 05].

Affective computing makes this workable at scale. AI assistants can infer emotional tone from brief signals (word choice, timing, prosody), select wording that fits the target state (reassuring vs. energizing) and adapt as patterns emerge. Early systems that explicitly model and respond to user affect (e.g. character-based companions that shift feedback when frustration rises) improve persistence compared to flat, one-size-fits-all prompts [PRE 05]. In applied settings, emotionally attuned conversational agents have delivered measurable gains in long-run adherence and self-management, suggesting that empathy-consistent phrasing and timing – not pressure – keep people returning to tasks [HOQ 13]. The through-line is modest and practical: assistants use emotion to lower friction and sustain momentum, leaving the choice set intact while making the next good action feel clear, doable and timely.

These examples only sketch the range of applications that are possible. In practice, emotionally attuned, AI-mediated nudges span far beyond these domains; wherever decisions hinge on attention, tone and timing, their careful use can convert intention into steady action.

7.5. Ethical and societal implications

7.5.1. *Autonomy and manipulation*

If we consider that someone profiles them and chooses the news they see, the news they do not see, the timing and the sequence, so as to steer interpretation, most people will find it unacceptable, almost akin to governance by design rather than well-intended assistance. This is what large-scale algorithmic systems already do. When the emotional layer is added, the influence deepens, with clear implications for *individual autonomy* and *democratic deliberation* [YEU 17], [LEA 25].

A thin but clear ethical line distinguishes supportive agency from manipulative engineering. Emotional cues can enlarge agency when they remove friction, clarify value and help people act on considered preferences. They cross into *manipulation* when they exploit transient states (such as fatigue, anxiety or elation) to drive actions that a minimally reflective person would not endorse, or when affect-tuned curation renders alternatives technically present but practically invisible [SUN 15], [FLO 22]. Hypernudging heightens manipulation risks by rapidly locating "emotional soft spots" and overfitting to them, narrowing exposure through many micro-adjustments that track mood and moment. In gatekeeper ecosystems, where the same stack measures, ranks and delivers, this can shift the horizon of reasons people routinely meet, undermining agency not by force but by accumulation.

Design should also anticipate *reactance* and *over-reliance* as ethical failure modes that directly affect the individual agency. Heavy or repetitive affective cues provoke reactance, a motivated resistance against perceived manipulation [BRE 66]. Conversely, endlessly comforting and ever-present assistance can foster excessive dependence, eroding self-regulation and practical autonomy in choice [SHA 20]. Light-touch, optional, pace-limited cues, punctuated by moments in which control is returned to the individual, help preserve agency while avoiding covert manipulation.

Consent and privacy also shape how emotional inference affects fairness. Generic or bundled consent cannot legitimize sensitive inferences from proxies such as wording, prosody or facial displays. Given weak evidential grounding and significant cross-domain risks, safeguards should include minimizing data collection, favoring on-device estimation, strict purpose binding, short retention periods, obtaining specific consent for any reuse and providing simple user controls, paired with stronger regulation of AI systems that claim to interpret human emotions. The EU AI Act – [EUR 24] restricts or prohibits certain emotion-inference uses and classifies others as high risk, providing a legal framework for protecting EU residents from embedded affect-aware AI.

7.5.2. *Bias, discrimination and fairness*

The difficulty begins with the construct itself. There is no single, uncontested taxonomy of emotion, and most systems infer affect from externalizations such as wording, prosody, facial displays and posture rather than from internal states. Treating these proxies as ground truth invites category error and overconfidence, which undermines fairness whenever proxy–state correlations are weak or unstable [BAR 17].

In addition to this conceptual fragility, operational weaknesses must also be considered. AI models are only as sound as their training datasets and inherit the *prejudices embedded in the training data.* Underrepresentation of dialects, accents, skin tones, age groups or interaction styles leads models to amplify historical skew at scale. The consequence is measurable bias in estimation and downstream discrimination, with clear losses of fairness for the most affected groups [BAR 23].

These vulnerabilities are magnified by *cultural variability* in expression repertoires and display rules. Even basic facial movements, when combined, are interpreted differently across cultures, resulting in an “in-group advantage” in recognition accuracy. Decoders tuned to one cultural pattern systematically misread others, turning measurement bias into operational discrimination once tone, pacing or salience are adapted based on those misreadings [ELF 02], [JAC 12].

Taken together, conceptual fragility, data skew and cultural variability create a straight channel from measurement bias to operational discrimination, with fairness eroding quietly in everyday interactions. Even under the banner of libertarian paternalism, affect-adaptive systems can inflict *potential harm*, particularly on vulnerable populations, when they act on unstable proxies and narrow the range of options that users systematically encounter.

7.5.3. *Guiding principles for responsible emotional nudging*

Emotional nudging operates at the most sensitive layer of influence: the way people feel as they make decisions. Because it both infers and elicits affect and can iterate in real time across individuals and contexts, clear guiding principles are needed to keep influence proportionate, auditable and aligned with human agency.

We propose a set of guiding principles for responsible emotional nudging. These principles are deliberately aligned with the emerging regulatory landscape, including the EU AI Act's treatment of affective computing and emotion–recognition systems as high-risk applications, so that emotional nudging remains both behaviorally effective and consistent with evolving requirements on transparency, accountability and fundamental rights.

The list that follows adapts and extends existing guiding principles of responsible digital nudging, as outlined by [LEA 25], to emotion-centric settings, tightening expectations around informed consent, transparency and the protection of vulnerable groups [WAC 17]. It does not aim for an exhaustive listing but a framework to guide debate and practice, aimed at safeguarding the ethical use of emotion-based mechanisms in policy.

The following guiding principles for responsible emotional nudging set a practical ethical baseline:

Beneficial intention

Aim interventions at clear, defensible welfare goals, with harms anticipated ex ante. Emotional cues should lower friction and support sound judgment rather than maximize engagement or revenue.

Autonomy

Preserve genuine freedom to choose and to refuse. Emotional modulation must not bypass reflective control; users should be able to mute, pause or reset the affective layer at any time.

Acceptance

Ensure the targeted audience views the intervention as appropriate and legitimate. Consider acceptability in context (tone, timing, frequency) and withdraw designs that trigger sustained discomfort or distrust, by the individual or by society.

Consent

Obtain specific, informed and revocable permission for collecting or inferring affective signals. Consent must cover purposes, data types, retention and reuse, with simple controls to allow for revisiting choices.

Privacy

Treat emotional data as high-risk and establish areas of unacceptable use. Minimize capture to what is strictly necessary, prefer on-device processing, bind the purpose tightly, set a short retention period and prohibit secondary use without fresh consent.

Respect for vulnerable groups

Identify populations at a greater risk of harm (e.g. minors, the elderly, people with disabilities or economically precarious users) and apply stricter thresholds, softer defaults and added safeguards.

Transparency

Make the presence and aims of emotional nudges visible in plain language. Users should know that affective signals influence delivery and be able to see "what is active" in the current interaction.

Explainability

Ensure that emotional nudges are not black boxes and are always explainable. AI nudging systems must operate within explainable AI (XAI), which provides clear and faithful reasons that link the affective inputs used to the nudge delivered. Provide concise "why this, why now" rationales that link observed cues to selected actions, comprehensible to non-experts.

Accountability

Keep traceable records of goals, data uses, model updates and policy changes. Enable internal oversight and independent audit, and offer practical channels for challenge, remedy and redress.

These guiding principles serve as a foundation, not a ceiling. In emotion-centric hypernudging, they should be operationalized into concrete design checks, data and policy audits, and governance routines that run continuously, not just at launch. Contexts will differ, but the aim remains constant: to maintain proportionate and auditable influence while preserving agency and trust. As affective technologies scale and mutate, the regulatory agenda grows more pressing and must be revisited frequently to keep pace with their rapid evolution.

Emotional nudging has a profound capacity to shape social relations. As nudging becomes more central and moves upstream, guiding attention and preference through affect, individual and collective agency will increasingly depend on how that power is bounded and regulated. The ethical choice is clear: channel affective AI systems to support agency and public reason or allow affective AI persuasion to set the terms of life in society.

7.6. Final remark: long-term impact on human behavior, trust and rationality

Emotion-sensitive nudging may smooth frictions in the short term, yet its long-term effects depend on how it reshapes habits of attention, commitment to choice and self-control. Systems that routinely calibrate emotional tone and timing may help people move forward, but they can also cultivate overreliance (outsourcing decision-making to AI systems), reactance (resistance to perceived guidance) or narrowed preference formation (learning within a restricted affective tunnel). These dynamics alter not only what is chosen today but also how preferences and heuristics evolve over the long term.

Trust in affective AI prompts depends on how they are implemented. When affective cues are transparent, proportionate and reversible, users learn that the guidance is helpful, calibrated and non-coercive, which strengthens institutional trust. When cues are opaque, repetitive or tailored exclusively for engagement, users learn to disregard interfaces, degrading their own signals and removing trust in these mechanisms. This is a costly balance for policies that present themselves as decision assistants to help users in various dimensions of daily life.

The long-term question is whether affection is channeled to broaden deliberation and agency or to capture attention and narrow decision-making. The answer will determine not only the trajectory of decision support based on big data processing and tailored interventions, but also the credibility of institutions and the quality of judgment in the coming years. The strategic challenge for policymakers and regulators is clear: *ensure trustworthy interactions* and *safeguard autonomy*, by setting guardrails that keep affective influence proportionate and auditable, so that affect is channeled to enlarge deliberation and agency rather than to harvest attention.

7.7. Abbreviations

AI – Artificial intelligence

EEG – Electroencephalography

ESG – Environmental, Social, and Governance

EU – European Union

LLM – Large Language Model

NLP – Natural Language Processing

XAI – Explainable artificial intelligence

7.8. Acknowledgments

This research has been financed by the National Funds of the Portuguese Foundation for Science and Technology (FCT), and, when eligible, co-financed by European funds, within the project UID/03182/2025, Centre for Research in Economics and Management, University of Minho (doi: 10.54499/UID/03182/ 2025).

7.9. Declaration

This chapter was edited with the aid of the AI-powered writing assistant Grammarly to enhance correctness and clarity.

7.10. References

[ALL 11] ALLCOTT H., "Social norms and energy conservation", *Journal of Public Economics*, vol. 95, nos 9–10, pp. 1082–1095, 2011.

[BAR 17] BARRETT L.F., *How Emotions are Made: The Secret Life of the Brain*, Pan Macmillan, London, 2017.

[BAR 23] BAROCAS S., HARDT M., NARAYANAN A., *Fairness and Machine Learning: Limitations and Opportunities*, MIT Press, Cambridge, MA, 2023.

[BIC 05] BICKMORE T., PICARD R.W., "Establishing and maintaining long-term human–computer relationships", *Interacting with Computers*, vol. 17, no. 4, pp. 711–736, 2005.

[BRE 66] BREHM J.W., *A Theory of Psychological Reactance*, Academic Press, New York, 1966.

[CAL 10] CALVO R.A., D'MELLO S., "Affect detection: An interdisciplinary review of models, methods, and their applications", *IEEE Transactions on Affective Computing*, vol. 1, no. 1, pp. 18–37, 2010.

[COW 17] COWEN A.S., KELTNER D., "Self-report captures 27 distinct categories of emotion bridged by continuous gradients", *Proceedings of the National Academy of Sciences*, vol. 114, no. 38, pp. E7900–E7909, 2017.

[DAM 03] DAMASIO A.R., *Looking for Spinoza: Joy, Sorrow, and the Feeling Brain*, Harcourt, New York, 2003.

[DAM 18] DAMASIO A.R., *The Strange Order of Things: Life, Feeling, and the Making of Cultures*, Pantheon, New York, 2018.

[DAR 72] DARWIN C., *The Expression of the Emotions in Man and Animals*, vol. 3, John Murray, London, 1872.

[DME 12] D'MELLO S., GRAESSER A., "Dynamics of affective states during complex learning", *Learning and Instruction*, vol. 22, no. 2, pp. 145–157, 2012.

[EKM 92] EKMAN P., "An argument for basic emotions", *Cognition & Emotion*, vol. 6, nos 3–4, pp. 169–200, 1992.

[ELF 02] ELFENBEIN H.A., AMBADY N., "On the universality and cultural specificity of emotion recognition: A meta-analysis", *Psychological Bulletin*, vol. 128, no. 2, pp. 203–235, 2002.

[EUR 24] EUROPEAN PARLIAMENT, EU AI Act: First regulation on artificial intelligence, Brussels, 2024.

[FER 16] FERRARA E., VAROL O., DAVIS C. et al., "The rise of social bots", *Communications of the ACM*, vol. 59, no. 7, pp. 96–104, 2016.

[FLO 22] FLORIDI L., COWLS J., "A unified framework of five principles for AI in society", in CARTA S. (ed.), *Machine Learning and the City: Applications in Architecture and Urban Design*, Wiley, Hoboken, NJ, pp. 535–545, 2022.

[GAS 24] GASTEIGER N., LIM J., HELLOU M. et al., "A scoping review of the literature on prosodic elements related to emotional speech in human-robot interaction", *International Journal of Social Robotics*, vol. 16, pp. 659–670, 2024.

[GEO 24] GEORGE S.M., ILYAS P.M., "A review on speech emotion recognition: A survey, recent advances, challenges, and the influence of noise", *Neurocomputing*, vol. 568, p. 127015, 2024.

[HO 23] HO M.T., MANTELLO P., HO M.T., "An analytical framework for studying attitude towards emotional AI: The three-pronged approach", *MethodsX*, vol. 10, p. 102149, 2023.

[HOQ 13] HOQUE M.E., COURGEON M., MARTIN J.-C. et al., "MACH: My automated conversation coach", in *Proceedings of the 2013 ACM International Joint Conference on Pervasive and Ubiquitous Computing (UbiComp '13)*, pp. 697–706, 2013.

[JAC 12] JACK R.E., GARROD O.G.B., YU H. et al., "Facial expressions of emotion are not culturally universal", *Proceedings of the National Academy of Sciences*, vol. 109, no. 19, pp. 7241–7244, 2012.

[JES 21] JESSE M., JANNACH D., "Digital nudging with recommender systems: Survey and future directions", *Computers in Human Behavior Reports*, vol. 3, p. 100052, 2021.

[KEL 10] KELTNER D., LERNER J.S., "Emotion", in FISKE S.T., GILBERT D.T., LINDZEY G. (eds), *Handbook of Social Psychology*, 5th edition, Wiley, Hoboken, NJ, pp. 317–352, 2010.

[LEA 21] LEAL C.C., OLIVEIRA B., "Choice architecture: Nudging for sustainable behavior", in MACHADO C. and DAVIM J.-P. (eds), *Sustainable Management for Managers and Engineers*, ISTE Ltd, London and Wiley, Hoboken, NJ, pp. 1–17, 2021.

[LEA 22] LEAL C.C., BRANCO-ILLODO I., OLIVEIRA B.M.N. et al., "Nudging and choice architecture: Perspectives and challenges", *Journal of Contemporary Administration*, vol. 26, no. 5 (Special Issue), e220098, 2022.

[LEA 24] LEAL C.C., OLIVEIRA B., "Nudging financial behavior in the age of artificial intelligence", in MACHADO C. and DAVIM J.-P. (eds), *Artificial Intelligence in Production Engineering and Management*, Elsevier, Oxford, pp. 115–144, 2024.

[LEA 25] LEAL C.C., OLIVEIRA B., "The ethics of digital nudging", in MACHADO C. (ed.), *Ethics in Management and Business*, Springer, Cham, pp. 53–80, 2025.

[LED 12] LEDOUX J., "Rethinking the emotional brain", *Neuron*, vol. 73, no. 4, pp. 653–676, 2012.

[LER 15] LERNER J.S., LI Y., VALDESOLO P. et al., "Emotion and decision making", *Annual Review of Psychology*, vol. 66, pp. 799–823, 2015.

[LOE 01] LOEWENSTEIN G.F., WEBER E.U., HSEE C.K. et al., "Risk as feelings", *Psychological Bulletin*, vol. 127, no. 2, pp. 267–286, 2001.

[MAD 01] MADRIAN B.C., SHEA D.F., "The power of suggestion: Inertia in 401(k) participation and savings behavior", *Quarterly Journal of Economics*, vol. 116, no. 4, pp. 1149–1187, 2001.

[MCS 18] MCSTAY A., *Emotional AI: The Rise of Empathic Media*, SAGE Publications, London, 2018.

[MIL 21] MILKMAN K.L., PATEL M.S., GANDHI L. et al., "A megastudy of text-based nudges encouraging patients to get vaccinated at an upcoming doctor's appointment", *Proceedings of the National Academy of Sciences*, vol. 118, no. 20, e2101165118, 2021.

[OLI 24] OLIVEIRA B., LEAL C.C., "Leading the way toward a sustainable future: The role of sustainable finance and environmental, social, and governance investing", in MACHADO C. and DAVIM J.-P. (eds), *Circular Economy and Manufacturing*, Elsevier, Oxford, pp. 53–82, 2024.

[OLI 25] OLIVEIRA B., LEAL C.C., "Artificial intelligence in finance: Applications and challenges", in MACHADO C. and DAVIM J.-P. (eds), *Challenges and Opportunities in Artificial Intelligence Era*, Springer, Cham, 2025.

[PAN 23] PAN B., HIROTA K., JIA Z. et al., "A review of multimodal emotion recognition: Datasets, preprocessing, features, and fusion methods", *Neurocomputing*, vol. 561, p. 126866, 2023.

[PAT 16] PATEL M.S., ASCH D.A., ROSIN R. et al., "Individual versus team-based financial incentives to increase physical activity: A randomized trial", *JAMA Internal Medicine*, vol. 176, no. 7, pp. 1007–1014, 2016.

[PEK 06] PEKRUN R., "The control-value theory of achievement emotions: Assumptions, corollaries, and implications for educational research and practice", *Educational Psychology Review*, vol. 18, no. 4, pp. 315–341, 2006.

[PIC 00] PICARD R.W., *Affective Computing*, MIT Press, Cambridge, MA, 2000.

[PIC 08] PICHERT D., KATSIKOPOULOS K.V., "Green defaults: Information presentation and pro-environmental behaviour", *Journal of Environmental Psychology*, vol. 28, no. 1, pp. 63–73, 2008.

[PRE 05] PRENDINGER H., ISHIZUKA M., "The empathic companion: A character-based interface that addresses users' affective states", *Applied Artificial Intelligence*, vol. 19, nos 3–4, pp. 267–285, 2005.

[SHA 20] SHAW P., "Building trust through sound governance", in CHISHTI S., BARTOLETTI I., LESLIE A., MILLIE S.M. (eds), *The AI Book: The Artificial Intelligence Handbook for Investors, Entrepreneurs and FinTech Visionaries*, Wiley, West Sussex, pp. 175–179, 2020.

[SPI 02] SPINOZA B., *Complete Works*, CURLEY E. (ed. and trans.), Hackett, Indianapolis, IN, 2002 [Original published 1677].

[SUN 15] SUNSTEIN C.R., *Choosing not to Choose: Understanding the Value of Choice*, Oxford University Press, Oxford, 2015.

[THA 04] THALER R.H., BENARTZI S., "Save more tomorrow: Using behavioral economics to increase employee saving", *Journal of Political Economy*, vol. 112, no. S1, pp. S164–S187, 2004.

[THA 08] THALER R.H., SUNSTEIN C.R., *Nudge: Improving Decisions about Health, Wealth, and Happiness*, Yale University Press, New Haven, 2008.

[WAC 17] WACHTER S., MITTELSTADT B., FLORIDI L., "Transparent, explainable, and accountable AI for robotics", *Science Robotics*, vol. 2, no. 6, eaan6080, 2017.

[WEI 16] WEINMANN M., SCHNEIDER C., VOM BROCKE J., "Digital nudging", *Business & Information Systems Engineering*, vol. 58, no. 6, pp. 433–436, 2016.

[WHI 14] WHITEHEAD M., JONES R., HOWELL R. et al., Nudging all over the world, ESRC Report, Economic and Social Research Council, Swindon and Edinburgh, 2014.

[YEU 17] YEUNG K., "Hypernudge: Big Data as a mode of regulation by design", *Information, Communication & Society*, vol. 20, no. 1, pp. 118–136, 2017.

8

Agentic AI in Marketing: Opportunities, Challenges and Impact on Firm Performance

The goal of this chapter is to explore the opportunities and challenges of AI agents and agentic artificial intelligence (AAI) systems adoption in marketing processes through the lens of Agency theory. We will identify specific applications and supporting tools for AAI systems in marketing and their potential impact on firm performance. We will conclude with several key points and managerial recommendations.

8.1. Introduction

Artificial intelligence (AI) is a rapidly evolving field that aims to understand and build intelligent entities capable of "compute how to act effectively and safely in a wide variety of novel situations" [RUS 22, p. 19]. Despite a history spanning more than seven decades, this topic has only recently attracted significant research effort, especially after 2022, with the commercial launch of Generative AI (GenAI) systems. Furthermore, the emergence of agentic artificial intelligence (AAI), characterized by autonomy, intentionality and collaborative abilities, represents the cornerstone of a new era in which AI technology is entering [HUG 25].

The rapid pace of technological development in the AAI area will transform industries, business processes, governance, working practices, consumer behavior and interactions among humans, humans and machines and among machines. For example, the launch of Manus AI, an agent with autonomous thinking, planning and execution capabilities, signals the entry into the era of AI as a workforce. Despite

Chapter written by Florin Sabin FOLTEAN and Octavian Dumitru HERA.

the unprecedented transformative impact of AAI, the topic of AI agents and AAI systems remains under-researched across academic fields, including marketing [HUG 25]. The goal of this chapter is to fill this knowledge gap by exploring the opportunities and challenges of AI agent and AAI system adoption in marketing processes through the lens of Agency theory.

In the remainder of this chapter, we will organize as follows. We will start by exploring the evolving concept of human and non-human agency in Agency theory. Based on this theoretical foundation, we will next clarify the conceptual domains of AI agents and AAI systems. We will continue to identify applications and supporting tools for AAI systems in marketing, as well as the challenges managers face in adopting them. Finally, we will explore the potential business value of AI agents and AAI systems. We will conclude with several key points and managerial recommendations.

8.2. AAI systems

The 2025 Technology Trends Outlook [MCK 25a] highlights AAI as a key focus area, capturing managerial interest and experimentation in both business and consumer technology. AAI combines the flexibility and generality of AI foundation models with the ability to act by creating "virtual coworkers" that can autonomously plan and execute multistep workflows.

8.2.1. *AI agency concept*

Interactions between people and technology have been studied through the lens of specific social theories, such as structural theory [GID 79, GID 84], practice theory [BOU 77], actor-network theory [LAT 87] and socio-materiality theory [ORL 07]. Particular research on human–technology interactions uses Agency theory as a theoretical foundation for understanding these complex relationships. [EMI 98] defined human agency in terms of "the temporally constructed engagement by actors of different structural environments – the temporal relational contexts of action – which, through the interplay of habit, imagination and judgment, both reproduces and transforms those structures in interactive response to the problems posed by changing historical situations" [EMI 98, p. 970].

In general, the term agency refers to "the capacity, condition, or state of acting or of exerting power" [MER 25]. The agency concept has been associated in sociological research with a variety of terms "selfhood, motivation, will, purposiveness, intentionality, choice, initiative, freedom, and creativity" [EMI 98, p. 962].

Agency theory, rooted in information economics, focuses on the "agency relationship, in which one party (the principal) delegates work to another (the agent), who performs that work" [EIS 89, p. 58]. Two key issues arise in agency relationships: monitoring the appropriateness of the agent's behavior and ensuring the alignment of the principal's and agent's preferences regarding specific actions and risks [EIS 89]. To resolve these problems, agency theorists seek "the most efficient contract governing the principal-agent relationship" [EIS 89, p. 58], focusing on expected behavior rather than expected outcomes.

Based on [ROS 73]'s argument for the universality of agency, this concept became relevant to both humans and non-human entities (e.g. AAI systems). Conceptualizing the agency of humans in terms of a temporally embedded process, [EMI 98] proposed three dimensions of this construct: (a) iteration ("selective reactivation by actors of past patterns of thought and action, as routinely incorporated in practical activity" [EMI 98, p. 971]); (b) projectivity ("the imaginative generation by actors of possible future trajectories of action" ([EMI 98, p. 971]); and (c) practical evaluation ("the capacity of actors to make practical and normative judgments among alternative possible trajectories of action" [EMI 98, p. 971]). Machine learning (ML) technologies and AI algorithms enhance AI systems in these dimensions: pattern recognition enables iteration, predictive models generate future action alternatives, and recommendation models evaluate alternative courses of action.

New AI technologies are transforming human–technology interactions and challenging the anthropocentric perspective on agency. [SCH 24] emphasizes the imperative, in the age of AAI, of rethinking human and machine agency and argues for the development of a post-anthropocentric theory of agency. [ZIE 98] proposes a conceptual distinction between behavioral autonomy and operational autonomy, which enables an understanding of the lower autonomy levels of rule-based approaches and the higher levels specific to AAI systems, which can make autonomous decisions and even act in unpredictable situations [SCH 24].

Attribution agency to AI systems should consider which kind of agency these systems can perform: phenomenological (i.e. sensing agency, feeling in control), means (degree of involvement in generating an outcome) and ends (capability of setting and achieving objectives) [VUK 17, cited in JON 25].

The extant information systems literature on AI agency treats the machine as an agent. It investigates the mechanics, dynamics and ethics of human(as principal)–machine(as agent) relationships through the lens of Agency theory [JON 25]. Previous research on human and material agency has adopted three alternative approaches: first, an asymmetric approach that privileges human actions; second, a symmetric approach that privileges neither human nor technological actions; and third, an approach that attributes agency to neither technology nor humans [DAT 23].

Previous studies on GenAI have focused on the capabilities, performance, autonomy and human-like behavior of these systems [JON 25]. Focusing on the agency idea of acting, [HUG 25] understood the agency of AI systems as "the ability to act" of these systems. The concept of AI (i.e. non-human) agency contributes to defining intelligent entities [e.g. RUS 22], emphasizing that it enables them to perform efficient and safe actions in novel situations.

[HUG 25] argued for conceptualizing the agency of AI systems as a multidimensional construct (i.e. a set of systems and human agencies). [DAT 23] emphasized the importance of distinguishing between the agency and the actions of traditional information systems. They proposed several dimensions of agency to facilitate this distinction: (a) passivity; (b) automaticity; (c) rationality; (d) endorsement; (e) freedom to choose; and (f) consciousness. The varying degrees of agency arise from the extent to which entities demonstrate these dimensions in their actions. AAI is deployed by both developers (who build, integrate and manage AI systems) and users (who deploy them for their own or their customers' use), and issues of delegation, collaboration, responsibility and accountability arise in both cases [HUG 25].

8.2.2. *AI agents*

The term "AI agent" was introduced in 1998 and describes a limited agent that lacked self-learning, generative reasoning and adaptability to dynamic environments [SAP 25]. [HUG 25] defined AI agents as "autonomous technologies, designed to perform tasks, make decisions and interact with their environment to achieve specific predefined goals, with minimal human intervention" [HUG 25, p. 5]. By integrating the capabilities of independent reasoning and action, AI agents exhibit autonomy and self-learning [HUG 25]. These represent the evolution from traditional computer software, enabling them to plan contextually in evolving environments [MIT 25]. Their autonomous decision-making, adaptability and collaborative abilities drive not only efficiency but also innovation and scalability [HUG 25].

The introduction of GenAI models enabled AI agents to generate coherent output. However, its operation remains relatively isolated from other agents [HUG 25] and access to real-time information reduces its potential to act strategically and efficiently [KIM 26]. Recent advances in technology enable AI agents to connect and thus to act autonomously, making goal-oriented decisions, adapting to dynamic environments and collaborating with other agents, whether AI or humans [HUG 25].

Provider	Label	Description
Microsoft	Website Q&A	It answers user questions and is deployed on the web, in Teams, or in other applications.
	Sales Qualification Agent	Evaluates each lead (background, recent opportunities, other data) based on a given set of criteria.
	Case Management Agent	The system automates key tasks throughout the case life cycle – creation, resolution, follow-up, closure – to reduce handling time and alleviate the burden on the service representative.
	Customer Knowledge Management Case Management Agent	Ensure knowledge articles are kept perpetually up-to-date by analyzing case notes, transcripts, summaries and other artifacts from human-assisted cases.
	Researcher	Gathers direct and indirect data and uses a multi-step research process to identify business opportunities, market trends or specific customer insights.
	Analyst	Processes raw data and uses a chain-of-thought reasoning to extract customer patterns or to create more accurate predictions.
Google	Gems	They are custom AI experts that can save highly detailed prompt instructions for repeatable tasks.
Open AI	ChatGPT Agent	Executes specific tasks as instructed by the user, such as navigating websites to collect information, connecting to other data sources, interpreting code and filling out forms. It will pause to ask for clarification when needed.
Anthropic	Claude Code sub-agents	They are specialized AI assistants that operate in isolated contexts, each handling specific development tasks. For example, the Interface Designer Agent designs user interfaces and API specifications, while the Data Scientist Agent analyzes data and prepares data visualization and reporting.
UiPath	Customer Feedback Agent	Automates the collection of feedback from customers.
	Social Media Strategist Agent	Compares the social media strategies of a company and its competitors.
	Product Recommendation Agent	Provides customer suggestions.
	Email Rewriter Agent	Enhances email drafts.

Table 8.1. *Instrumental agents (source: the authors)*

Provider	Label	Description
Mistral	Financial Analyst	An advisory agent is permanently connected to multiple sources of financial data, compiling insights and recommendations.
Lindy	Inbound Sales Agent	Manages inbound sales, provides knowledge base responses, schedules follow-up calls and keeps the team informed about the progress.
	Sales Lead Scoring Agent	The system enriches lead data, calculates lead scores and notifies the team for high-priority leads.
	Inbound Sales Agent	Manages inbound sales, provides knowledge base responses, schedules follow-up calls and keeps the team informed about the progress.
Salesforce	Einstein	Engages with inbound prospects, makes decisions, prioritizes actions, analyzes context and autonomously determines the next steps.
Optimizely	Opal	Performs various marketing functions, such as content generation, campaign planning and maintaining brand guidelines while creating new ideas in a short time.
Jasper	Personalization Agent	The system adapts content to each segment without requiring specific prompting.
Relevance AI	Lifecycle Marketing Agent	Delivers a personalized message, tailored for every customer.
HubSpot	Content Agent	The system resolves inquiries that it can handle independently and escalates only important cases, allowing the team to save time.
	Social Media Agent	Creates social posts and suggests a publish calendar.
Lyzr	Skott	Automates and personalizes every step of the marketing funnel across content, campaigns, SEO and analytics.
Akira	Customer Profitability Analysis Agent	Automates data integration, builds segments based on customers' interactions, purchase profiles, profit levels and activities, and develops profitability models to increase revenue and lifetime value or decrease churn rate and service costs.
ZBrain	Content Research Agent	Creates new content after identifying relevant keywords and gathering web insights for the targeted audience.
	Brand Voice Analyzer Agent	Evaluates content to determine its tone, style and personality traits, helping to align messaging with brand identity.
	Fact Checking Agent	Ensures marketing content accuracy by verifying data, enhancing credibility and maintaining brand trustworthiness.
	Social Media Content Generator Agent	Creates new social media-tailored posts from user-uploaded content to drive higher engagement.

Table 8.2. *Collaborative agents (source: the authors)*

AI agents developed by leading companies divide into two categories: (a) instrumental and (b) collaborative. Instrumental AI agents are general-purpose, task-based agents that activate on demand to automate recurring tasks or assist with occasional tasks. Table 8.1 provides short descriptions of several instrumental agents.

Collaborative AI agents are specialized, ready-to-use agents that act as the workforce, integrate into business processes and collaborate with human employees. Table 8.2 provides short descriptions of several collaborative agents.

8.2.3. *Architecture and operational mechanisms of AAI systems*

AAI systems operate autonomously to achieve predefined objectives through iterative planning, task decomposition and continuous learning [BOR 25], without requiring human intervention over extended timeframes [ACH 25]. The distinctive characteristic of AAI systems lies in their ability to operate in real-world contexts, using digital tools to pursue specific goals rather than produce simple outputs [MCK 25b]. Due to the emergence of large language models (LLMs), the AAI systems augmented their ability to interact in a specific context [BOR 25], either with humans (in hybrid systems) or between AI agents exclusively, who work collectively to complete their mission in a more effective way than a single agent [PAR 24].

To enhance performance, AAI systems combine neural networks and symbolic AI frameworks, thereby improving their capacity for reasoning and planning [COL 25]. While neural networks learn from large datasets [NAW 25], symbolic AI uses rule-based and logic-driven representations [COL 25] to enhance knowledge from deductive reasoning [NAW 25]. The resulting framework, known as "Neuro-symbolic AI", exhibits superior performance, explainability and generalization [OTT 23]. Thus, agents can learn from human interactions using neural networks [NAW 25], consequently enhancing the AAI system's ability to operate autonomously in dynamic conditions [HUG 25].

GenAI has elevated the potential of AAI systems through natural language processing, generation and reasoning capabilities, which facilitate effective collaboration with human teams and improve the quality of decision-making [HUG 25].

Modern AAI systems can mirror the current setup of human organizations to integrate and function properly alongside human agents. Previous research suggests integrating artificial agents into groups of specialized agents that can collaborate to plan and autonomously allocate tasks, thereby completing the assignment [HER 24]. In the foreseeable future, we can expect further changes to AAI systems that better align with hybrid organizations where AI will be widespread. In this context, the AAI system may act as collective intelligence, guided by humans in setting strategic objectives but operating through multiple agents to retrieve data, plan and execute actions [KIM 26].

An AAI system architecture follows natural evolutionary paths. Natural cognitive architectures provide valuable insights into how we can organize AAI systems. Essentially, there are three main components: an external component to connect with the environment, a coding mechanism and an internal brain to process information [BOR 25]. The external component is akin to natural senses, tasked with collecting data from the outside [MUR 25]. Coding requires specialized agents who can translate this data into formats required for internal processing by additional specialized agents, which analyze alternatives and make optimal decisions [PAR 24]. Following this approach, [MUR 25] proposed an AAI system architecture composed of several interconnected components, such as perceptions, cognitive processing and decision-making, action execution and continuous learning, while [BOR 25] identified several classes of agents assigned with roles in this kind of setup, such as executive agents (who implement), observer agents (who decode user interactions) and surface agents (who deliver and capture user interactions). These agents are hierarchically organized and collaborate to achieve their goals through a process known as "orchestration" [SAP 25].

AAI systems have a modular structure that enables them to perform multiple tasks simultaneously, thereby achieving a goal. In contrast, traditional software programs typically focus on a single task [ACH 25]. The process of performing multiple tasks to achieve a goal is called "chaining" and enhances the autonomy of the AAI system [MUR 25].

While assessing what leads to successful implementation, [MCK 25c] found that, despite the apparent importance of AI agents' capabilities, the most important focus should be on the workflows within the AAI system and on how it can adapt to changing environments in which it operates. This approach would allow AI teams to deploy the right technology at the right time, starting with user pain points [MCK 25c].

Keeping track of prior experience also contributes to the AAI systems' long-term stability by leveraging memory models that retain contextual information and enable learning from past interactions to optimize subsequent actions [ACH 25]. In time, AAI systems can be automatically improved by monitoring and learning from their output. The process is called "Agentic Loop" and consists of continuous evaluation performed by AI agents [SAP 25]. [LEO 25] identified five phases of agentic loop stages: (1) delegation, where humans retain all authority. This designs and deploys AI as a tool; (2) attribution, a gradual transfer where AAI system gets a certain degree of autonomy; (3) contingency, where AAI system becomes limited and unable to tackle unexpected events; (4) reassertion, a reversal to human authority, enforcing its responsibility; and (5) reconfiguration, with adjusted systems for future interactions.

Each module of the AAI system is specialized and continually optimized. Its modularity also provides flexibility and scalability, allowing for the adaptation to change, including the reconfiguration of its modules [ACH 25].

At the same time, AAI systems require control mechanisms, such as retrieval-augmented generation (RAG), to minimize hallucination by retrieving knowledge from external sources and verifying it against the initial output [SAP 25]. Empowered with RAG abilities, the AAI system can be deployed with greater confidence as a conversational system capable of making real-time decisions [ACH 25]. While RAG provides control mechanisms through external sources, mutual reflexivity is a form of control that operates within the system, enabling agents to review each other's results, thereby contributing to the overall realism and quality of the system's output [SAP 25].

8.3. AAI systems opportunities in marketing

In marketing, AAI systems represent an evolution from predictive AI, which can effectively analyze and forecast, and GenAI, which can be used for creating roles [KSH 25]. Since AAI systems possess enhanced autonomy, they can be deployed effectively for ongoing monitoring and rapid action [MUR 25]. Moreover, it can be efficiently deployed in machine-to-machine interaction. For example, AAI systems can act autonomously based on extensive data collected from interconnected devices such as sensors, cameras and other sources [LU 20]. Integrating AAI systems into the edge computing layer optimizes processing time and ensures optimal response time in the Internet of Things (IoT) context [NAW 25].

The autonomy of AAI systems allows ongoing scaling and optimization across multiple digital touchpoints. They will use specialized agents for specific marketing tasks (data analysis, copywriting, email marketing, etc.), thereby mutually enhancing performance to increase engagement and drive growth [KSH 25].

8.3.1. *Applications of AAI systems in marketing processes*

Integrating AAI systems into marketing practices is still in its early stages, with marketing technology companies seeking specific ways to use AAI systems within existing marketing processes. In marketing, AAI systems can have multiple applications. For example, they can serve as substitutes for a large team of human marketing specialists, handling various personalized interactions simultaneously

[PAR 24]. As a result, they will achieve superior engagement and user satisfaction through their instant responses and personalized, high-quality decision-making [MUR 25].

AAI systems can also be assigned to explore new market opportunities and manage campaigns, while generating new agents for specific tasks when required [KSH 25]. For example, a market research company deployed an AAI system to identify data anomalies and explain changes in market performance, achieving a 60% increase in productivity to complete the assigned task [MCK 25b]. Another way to increase productivity in market research by reducing costs and time-to-execution is to use synthetic research, which uses agents that replicate human profiles rather than relying on human participants [HER 24].

Several other use cases of deployed AAI systems in marketing are already visible on the market. For example:

– *Content marketing*: EmpathyLab can instantly generate dynamic, personalized content based on micro-profiling and real-time variables, and proactively deliver actionable insights by monitoring competition and social trends (EmpathyLab website).

– *Marketing campaigns*: Artefact autonomously optimizes campaign performance and makes real-time adjustments based on multiple performance variables (Artefact website). Qualcomm introduces an AAI system that autonomously creates personalized campaign kits with tailored key messages, value pillars and proof points based on web searches (Writer website).

– *Customer journey*: Braze uses AAI to personalize every customer journey based on unique customer actions, behaviors and pattern shifts, thereby improving conversion rates (Braze website).

– *CRM platforms*: Salesforce has introduced Agentforce, a platform that enables businesses to build and customize autonomous AI agents (Sirocco website). These agents can connect to any enterprise data and operate across sales, service, marketing and commerce. Microsoft has integrated AI across its Dynamics 365 platform, supporting marketing, sales and customer support functions by generating content, surfacing insights and summarizing customer interactions (Microsoft website).

8.3.2. *Architecture of AAI systems in marketing*

In the agentic economy, a marketing AAI system can perform similarly to a marketing department, executing its functions with agent roles such as data analyst, content creator or social media admin. This structure enables continuous scaling and optimization, with each agent enhancing the capabilities of others, resulting in highly efficient, dynamic operations [SAP 25].

According to [ERN 25], performant agentic infrastructure would include several workflows: sequential workflow (predefined specialization), parallel workflow (independent, complementary tasks), hierarchical workflow (categorized into groups of supervisors and subordinates) and iterative workflow (for ongoing adjustments and optimization). Optimizing tasks becomes more effective with multi-agent systems, as each agent is specifically designed for its purpose.

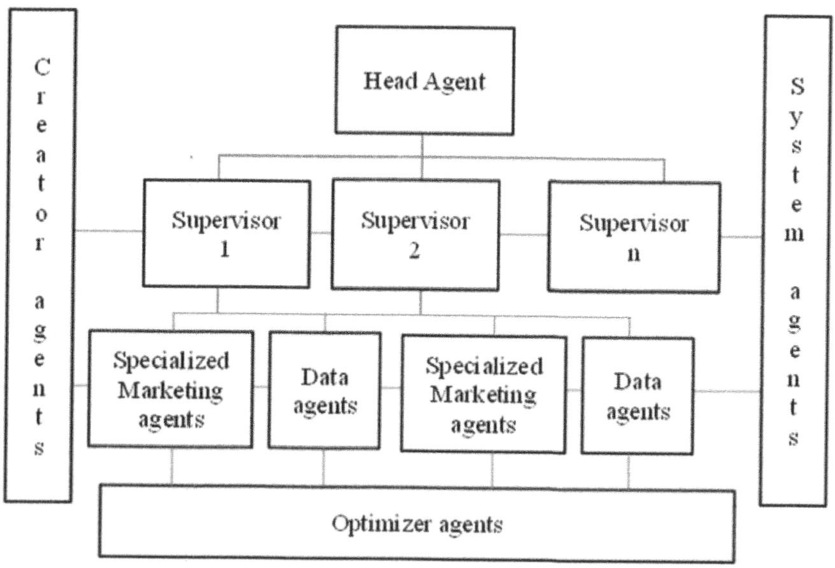

Figure 8.1. *Marketing AAI system architecture (source: the authors)*

The main building blocks of the marketing AAI system architecture are: (a) head agent; (b) supervisor agents; (c) marketing agents; (d) data agents; (e) system agents; (f) creator agents; and (g) optimizer agents. The head agent directs the agent system to achieve the business's goals and collaborates with human agents. Supervisor agents orchestrate agents that plan and divide tasks among teams of specialized agents. Marketing agents handle various tasks autonomously, including campaign planning and content creation. System agents protect the system internally (integrity) and externally (security), track and audit actions, and provide explainable input for control. Data agents specialize in gathering required information and connecting with other systems or platforms to capture necessary data. Creator agents create new agents or reconfigure existing ones as needed. Optimizer agents capture feedback, measure performance and contribute to the optimization of the system.

To protect the security of their AAI systems against cyberattacks or jailbreaking, companies may also consider building a shadow system as a control mechanism that runs in parallel with the actual system, assesses its proper behavior and highlights any divergences when they occur [HER 24].

8.4. Challenges of AAI systems adoption in marketing organizations

Managers face several challenges in their journey to integrate agent systems into existing marketing organizations. Among the most critical challenges facing AAI systems, [PAR 24] identified bias, unintended consequences and accountability. Starting from these critical challenges, we have compiled a more comprehensive list of concerns regarding the adoption of AAI systems in marketing.

– *Bias amplification effect*: since AI agents in the AAI system have been individually trained on historical datasets, they may inherit biases that could propagate and amplify within the system, leading to greater inequities than in isolated models [SAP 25]. To mitigate bias amplification, AAI implementations should make additional efforts to obtain high-quality training datasets [ACH 25].

– *Accountability issue*: as an AAI system begins to operate, it may occasionally exhibit unintended behavior. This behavior may be caused by LLMs' tendency to hallucinate, a lack of causal understanding (agents' limitations in understanding cause-and-effect relations) or divergent tasks and insufficient goal alignment [SAP 25], requiring constant monitoring by human operators [MCK 25a]. Furthermore, legal frameworks that address accountability will play a crucial role in ensuring trust in these systems [LI 24].

– *AAI system explainability*: managers expect to understand how AI agents make decisions autonomously, anticipating the consequences for governance mechanisms, employee behavior and ethical practices [HUG 25]. Although AI agents are effective at detecting statistical correlations, they may lack a causal understanding, i.e. they may not comprehend the underlying cause of specific effects [SAP 25]. AAI systems are vulnerable to malicious actions that, if successfully executed, may compromise one agent and spread within the system when agents interact, manipulating their behavior [SAP 25]. In this respect, it is essential to have AAI systems with explainability to provide insights into how decisions are made [NAW 25], so that the causes of unintended consequences can be identified and corrected. However, the explainability of AAI systems is more complex, since each agent's causal chain is influenced by its own memory, task objective and reasoning path [SAP 25], making it more challenging to identify the causes of unintended consequences. A higher degree of explainability can be achieved through interactive dashboards or graphical interfaces and will positively influence understanding and trust in the AAI system [NAW 25]. Explainability can provide greater transparency into the processes performed by the AAI system. Still, the accountability for consequences remains unclear [BOR 25], as does the issue of authorship and intellectual property [HUG 25].

– *Human–AI collaboration*: as AAI systems are adopted, they will work alongside teams of human employees, posing new challenges in deploying hybrid teams. Dynamic hybrid team composition and human–AI agent collaboration are essential topics that require further research [BOR 25]. When AAI systems automate tasks and processes, human workforce substitution necessitates reskilling and lifelong learning in response to job displacement [MUR 25].

– *Social ability*: communication capabilities and social ability to adapt and resolve real-life interactions in the context of ambiguity, as exhibited by AI agents, facilitate effective collaboration with humans in hybrid teams [SAP 25].

– *Data quality and scarcity*: an AAI system would require substantial quantities of labeled data to achieve optimal performance, but in the real world, this may be limited or too costly [NAW 25]. Current technological shortcomings limit the adoption of AAI systems. For example, high-quality data needed to train performant specialized agents is still too costly to gather [NAW 25], while integration issues with existing IT systems create additional constraints [ACH 25]. AAI systems promise not only to be more potent than isolated agents based on large models but also more cost-effective, requiring less computing power and energy to train [HUG 25].

– *Limited agency*: despite the vast potential to perform more complex tasks currently handled by skilled human employees, many AI agents exhibit insufficient agency. Most AI agents require explicit user instructions to act, lacking the capacity to self-initiate task generation and auto-error correction [SAP 25].

– *Privacy issue and over-regulation*: the inherent opacity of AAI systems complicates compliance with regulations [HUG 25]. Progress in AAI systems may be weakened by EU overregulation in the AI field, adding extra barriers to innovation in the tech sector and thwarting the long-term competitiveness of European companies [WIL 25].

– *Safety guardrails*: AAI developers should design and implement tools and procedures that can intervene and correct AAI systems when their actions, either intentional or not, could cause harm to humans [RAP 25].

– *Ethical issue*: this issue arises when an AAI system becomes a moral agent, meaning it can make free choices, deliberate on moral reasons and act on those deliberations [TEL 25]. The current technological level of AAI systems is approaching the threshold of becoming a moral agent. Therefore, they should have a built-in moral system encompassing human values to ensure its alignment with our society. However, because artificial reasoning differs from human reasoning, creating an ethical, AAI system is extremely difficult [RIE 25]. [RUS 19] envisions a set of core principles that all AAI systems should possess: maximizing the realization of human values; acknowledging initial uncertainty about what humans value; and learning about human values by observing the choices humans make.

8.5. Business value of AAI systems in marketing

Recent technological advances in ML, GenAI and AAI systems, as well as big data, computing power and AI infrastructure, increase the potential of AAI to create business value and sustain a firm's competitive advantage [GRO 18, HUA 21, KRA 23]. Managers focus on the AAI business value, specifically on the mechanisms for achieving business growth and efficiency gains in the new AI era. The complexity of GenAI-driven AAI systems and the new value-creation opportunities they open capture managers' attention [QUE 24]. Consequently, recent literature has begun to explore the benefits and performance results of AI adoption and use [BAI 21].

AAI can transform value-creation and capture processes through features such as autonomous decision-making, intentional and purposeful action and real-time communication, thereby driving innovation and increasing process efficiency. AAI systems can capture consumer behavior data, enabling real-time sensing and the rapid capture of new marketing opportunities. The natural language processing and real-time communication capabilities of AI agents make the customer relationship management and customer support process more effective and efficient. As AI investments can generate business value and create new competitive advantages, companies have begun to allocate significant resources to AI technologies, including ML, GenAI, AAI and robotic process automation.

Managers can follow two main ways of using AI technologies in value creation processes: (a) human capabilities augmentation; and (b) process automation that substitutes for human capabilities [QUE 24]. On the one hand, human capabilities augmentation involves supporting human agents in information processing and decision-making by providing analysis, predictions and recommendations. On the other hand, in process automation, human agents delegate to the AAI system mostly (or entirely) the action rights.

The benefits managers expect from AI adoption and use include gains in profitability, operational efficiency and return on marketing-related investment [MIS 22]. Despite these potential AI benefits, ranging from improved decision-making to sustained business competitiveness and performance, managers hesitate to invest in new AI technologies because of the low level of explainability in the system's decision-making [FER 22]. Furthermore, they consider in the process of AI investment appraisal their delegation preferences and the consequences of delegating decision-making and action rights to AI [QUE 24].

Although business firms' adoption of AI has increased over the last few years, it remains relatively low. Obtaining superior benefits from AI investments relative to their costs requires a high level of organizational readiness for the adoption and effective use of these technologies.

8.6. Conclusion

In this chapter, we explored the concept of human and non-human agency in the new AAI context through the lens of Agency theory, as a theoretical foundation for discussing the conceptual scope of AI agents and AAI systems. In addition, we proposed an architecture for an AAI system and identified its applications in marketing. Finally, we examined the challenges managers encounter when implementing AAI systems in marketing and the potential business benefits of AAI.

Future research could focus on specific issues relevant to marketing managers, such as:

– How should organizational culture and structure be adapted to encourage the adoption and use of AAI in marketing processes?

– How could members of the marketing organization be incentivized and trained to collaborate with non-human agents?

– How will a firm's focus on AI and human–AI agent collaboration capabilities in marketing improve company performance?

As the world races to catch up in AI, the key question that remains is the long-term sustainability of AAI systems. Since we are yet in the exploration and early-adoption stage, managers accept the implementation and operational costs of AAI systems as part of an investment that promises a higher return in the future. On the contrary, as AAI systems mature, the focus will shift to efficiency. Careful planning and hybridization of marketing roles (human and AI agent) would ensure superior effectiveness in the future and would empower marketing teams to handle unforeseen conditions.

In conclusion, providing more guidance for marketing managers on adopting and utilizing AAI tools to enhance marketing capabilities and increase effectiveness will increase the likelihood that the firm will maintain a competitive advantage and superior performance.

8.7. References

[ACH 25] ACHARYA D.B., KUPPAN K., DIVYA B., "Agentic AI: Autonomous intelligence for complex goals: A comprehensive survey", *IEEE Access*, vol. 13, pp. 18912–18936, doi: 10.1109/ACCESS.2025.3532853, 2025.

[BAI 21] BAIRD A., MARUPING L., "The next generation of research on IS use: A theoretical framework of delegation to and from agentic IS artefacts", *MIS Quarterly*, vol. 45, no. 1, pp. 315–341, 2021.

[BOR 25] BORGHOFF U.M., BOTTONI P., PARESCHI R., "Human-artificial interaction in the age of agentic AI: A system-theoretical approach", *Frontiers in Human Dynamics*, vol. 7, pp. 1–16, doi: 10.3389/fhumd.2025.1579166, 2025.

[BOU 77] BOURDIEU P., *Outline of a Theory of Practice*, Cambridge University Press, Cambridge, 1977.

[COL 25] COLELOUGH B., REGLI W., "Neuro-Symbolic AI in 2024: A systematic review", *ArXiv*, doi: 2501.05435, 2025.

[DAT 23] DATTATHRANI S., DE' R., "The concept of agency in the era of artificial intelligence: dimensions and degrees", *Information Systems Frontiers*, vol. 25, no. 1, pp. 29–54, 2023.

[DUI 25] DUIVENVOORDE B., "Generative AI and the future of marketing: A consumer protection perspective", *Computer Law & Security Review*, vol. 57, pp. 1–14, doi: 10.1016/j.clsr.2025.106141, 2025.

[EIS 89] EISENHARDT K.M., "Agency theory: An assessment and review", *The Academy of Management Review*, vol. 14, no. 1, pp. 57–74, 1989.

[EMI 98] EMIRBAYER M., MISCHE A., "What is agency?", *American Journal of Sociology*, vol. 103, no. 4, pp. 962–1023, 1998.

[ERN 25] ERNST & YOUNG, Agentic AI. The next frontier in automation, available at: https://www.ey.com/en_in/insights/ai/agentic-ai-the-next-frontier-in-automation, 2025.

[FER 22] FERNÁNDEZ-LORÍA C., PROVOST F., HAN X., "Explaining data-driven decisions made by AI systems: The counterfactual approach", *MIS Quarterly*, vol. 46, no. 3, pp. 1635–1660, 2022.

[GID 79] GIDDENS A., *Central Problems in Social Theory*, Macmillan, London, 1979.

[GID 84] GIDDENS A., *The Constitution of Society: Outline of the Theory of Structuration*, Polity, Cambridge, 1984.

[GRO 18] GROVER V., CHIANG R.H., LIANG T.-P. et al., "Creating strategic business value from big data analytics: A research framework", *Journal of Management Information Systems*, vol. 35, no. 2, pp. 388–423, 2018.

[HER 24] HERMANN E., PUNTONI S., "Generative AI in marketing and principles for ethical design and deployment", *Journal of Public Policy & Marketing*, vol. 44, no. 3, pp. 332–349, doi: 10.1177/07439156241309874, 2024.

[HUA 21] HUANG M.-H., RUST R.T., "A strategic framework for artificial intelligence in marketing", *Journal of the Academy of Marketing Science*, vol. 49, pp. 30–50, 2021.

[HUG 25] HUGHES L., DWIVEDI Y.K., MALIK T. et al., "AI agents and agentic systems: A multi-expert analysis", *Journal of Computer Information Systems*, pp. 1–29, doi: 10.1080/08874417.2025.2483832, 2025.

[JON 25] JONES M., "I gotta use words when I talk to you: Primed suspension of disbelief in views on agency in relation to Artificial Intelligence", *Information and Organization*, vol. 35, p. 100557, doi: 10.1016/j.infoandorg.2025.100557, 2025.

[KIM 26] KIM S., HA J., LEE H. et al., "Human-guided collective LLM intelligence for strategic planning via two-stage information retrieval", *Information Processing & Management*, vol. 63, no. 1, pp. 1–33, doi: 10.1016/j.ipm.2025.104288, 2026.

[KRA 23] KRAKOWSKI S., LUGER J., RAISCH S., "Artificial intelligence and the changing sources of competitive advantage", *Strategic Management Journal*, vol. 44, no. 6, pp. 1425–1452, 2023.

[KSH 25] KSHETRI N., "From predictive and generative to agentic AI: Shaping the future of marketing operations and strategies", *Computer*, vol. 58, no. 4, pp. 121–129, doi: 10.1109/MC.2025.3530304, 2025.

[LAT 87] LATOUR B., *Science in Action: How to Follow Scientists and Engineers through Society*, Harvard University Press, Cambridge, MA, 1987.

[LEO 25] LEONARDI P.M., "Homo agenticus in the age of agentic AI: Agency loops, power displacement, and the circulation of responsibility", *Information and Organization*, vol. 35, no. 3, pp. 1–18, doi: 10.1016/j.infoandorg.2025.100582, 2025.

[LI 24] LI Y., WU B., HUANG Y. et al., "Developing trustworthy artificial intelligence: Insights from research on interpersonal, human-automation, and human-AI trust", *Frontiers in Psychology*, pp. 1–13, doi: 10.3389/fpsyg.2024.1382693, 2024.

[LU 20] LU Y., LIU C., WANG K.I.-K. et al., "Digital twin-driven smart manufacturing: Connotation, reference model, applications and research issues", *Robotics and Computer-Integrated Manufacturing*, vol. 61, pp. 1–14, doi: 10.1016/j.rcim.2019.101837, 2020.

[MCK 25a] MCKINSEY, Technology trends outlook 2025, fifth edition, Report, Mckinsey, 2025.

[MCK 25b] MCKINSEY, Seizing the agentic AI advantage, Report, Mckinsey, 2025.

[MCK 25c] MCKINSEY, One year of agentic AI: Six lessons from the people doing the work, Report, Mckinsey, 2025.

[MER 25] MERRIAM-WEBSTER DICTIONARY, available at: meriam-webster.com, 2025.

[MIS 22] MISHRA S., EWING M.T., COOPER H.B., "Artificial intelligence focus and firm performance", *Journal of the Academy of Marketing Science*, vol. 50, pp. 1176–1197, 2022.

[MIT 25] MITCHELL M., AVIJIT G., SASHA LUCCIONI A. et al., "Fully autonomous AI agents should not be developed", *ArXiv*, doi: 10.48550/arXiv.2502.02649, 2025.

[MUR 25] MURUGESAN S., "The rise of agentic AI: Implications, concerns, and the path forward", *IEEE Intelligent Systems*, vol. 40, no. 2, pp. 8–14, doi: 10.1109/MIS.2025.3544940, 2025.

[NAW 25] NAWAZ U., ANEES-UR-RAHAMAN M., SAEED Z., "A review of neuro-symbolic AI integrating reasoning and learning for advanced cognitive systems", *Intelligent Systems with Applications*, vol. 26, pp. 1–14, doi: 10.1016/j.iswa.2025.200541, 2025.

[ORL 07] ORLIKOWSKI W.J., "Sociomaterial practices: Exploring technology at work", *Organization Studies*, vol. 28, no. 9, pp. 1435–1448, 2007.

[OTT 23] OTT J., LEDAGUENEL A., HUDELOT C. et al., "How to think about benchmarking neurosymbolic AI?", in *7th International Workshop on Neural-Symbolic Learning and Reasoning -NESY 2023*, Siena, 2023.

[PAR 24] PARESCHI R., "Beyond human and machine: An architecture and methodology guideline for centaurian design", *Science*, vol. 6, no. 4, pp. 1–20, doi: 10.3390/sci6040071, 2024.

[QUE 24] QUEIROZ M., ANAND A., BAIRD A., "Manager appraisal of artificial intelligence investments", *Journal of Management Information Systems*, vol. 41, no. 3, pp. 682–707, doi: 10.1080/07421222.2024.2376383, 2024.

[RAP 25] RAPTIS E.K., KAPOUTSIS A.C., KOSMATOPOULOS E.B., "Agentic LLM-based robotic systems for real-world applications: A review on their agenticness and ethics", *Frontiers in Robotics and AI*, pp. 1–25, doi: 10.3389/frobt.2025.1605405, 2025.

[RIE 25] RIESEN E., "A sociotechnological-system approach to AI ethics", *AI & Society*, pp. 1–13, doi: 10.1007/s00146-025-02335-5, 2025.

[ROS 73] ROSS S.A., "The economic theory of agency: The principal's problem", *The American Economic Review*, vol. 63, no. 2, pp. 134–139, available at: http://www.jstor.org/stable/1817064, 1973.

[RUS 19] RUSSELL S., "Should we fear supersmart robots?", *Scientific American*, available at: https://www.scientificamerican.com/article/should-we-fear-supersmart-robots/, 2019.

[RUS 22] RUSSELL S.J., NORVIG P., *Artificial Intelligence: A Modern Approach*, 4th edition, Global Edition, Pearson Education Limited, Harlow, 2022.

[SAP 25] SAPKOTA R., ROUMELIOTIS K.I., KARKEE M., "AI agents vs. agentic AI: A conceptual taxonomy, applications and challenges", *Information Fusion*, vol. 126, no. B, pp. 1–30, doi: 10.1016/j.inffus.2025.103599, 2025.

[SCH 24] SCHREIBER G., "Reconsidering agency in the age of AI", *Filozofia*, vol. 79, no. 5, pp. 529–537, 2024.

[TEL 25] TELAKIVI P., KOKKONEN T., HAKLI R. et al., "AI-extended moral agency?", *Social Epistemology*, pp. 1–13, doi: 10.1080/02691728.2025.2472759, 2025.

[VUK 17] VUKOV J.M., "Three kinds of agency and closed-loop neural devices", *AJOB Neuroscience*, vol. 8, no. 2, pp. 90–91, 2017.

[WIL 25] WILKINSON I., "The EU's new AI code of practice has its critics but will be valuable for global governance", *Chatham House*, available at: https://www.almendron.com/tribuna/the-eus-new-ai-code-of-practice-has-its-critics-but-will-be-valuable-for-global-governance/, 2025.

[ZIE 98] ZIEMKE T., "Adaptive behavior in autonomous agents", *Presence: Teleoperators and Virtual Environments*, vol. 7, no. 6, 564–587, 1998.

List of websites

Akira website, https://www.akira.ai/ai-agents

Artefact website, https://www.artefact.com/blog/agentic-ai-in-practice-strategic-value-for-digital-and-data-marketers/

Braze website, https://www.braze.com/product/journey-orchestration

EmpathyLab website, https://www.empathylab.com/en/news-and-insights/agentic-AI-reshaping-marketing-team

Google website, https://gemini.google/overview/gems/

Hubspot website, https://www.hubspot.com/products/artificial-intelligence/breeze-ai-agents

Jasper website, https://www.jasper.ai/agents

Lindy website, https://www.lindy.ai/agents/sales-agents

Lyzr website, https://www.lyzr.ai/skott/

Microsoft website, https://www.microsoft.com/en-us/microsoft-copilot/blog/copilot-studio/explore-agents-pre-built-for-you-in-microsoft-copilot-studio/

Microsoft website, https://www.microsoft.com/en-us/microsoft-365/blog/2025/03/25/introducing-researcher-and-analyst-in-microsoft-365-copilot/

Microsoft website, https://www.microsoft.com/en-us/dynamics-365/blog/business-leader/2025/06/04/ai-first-crm-systems-learnings-from-organizations-making-the-switch/

OpenAI website, https://help.openai.com/en/articles/11752874-chatgpt-agent

Optimizely website, https://www.optimizely.com/insights/blog/ai-agents/

Relevance AI website, https://relevanceai.com/lima-lifecycle-marketing-agent

Salesforce website, https://www.salesforce.com/news/stories/einstein-sales-agents-announcement/

Sirocco website, https://www.siroccogroup.com/salesforce-agentforce/

Superprompt website, https://superprompt.com/blog/best-claude-code-agents-and-use-cases

UiPath website, https://www.uipath.com/solutions/technology/specialized-agents

Writer website, https://writer.com/blog/qualcomm-customer-story/

ZBrain website, https://zbrain.ai/agents/Marketing

9

AI's Role in Marketing: Mapping the Evolution of Creativity

Creativity has always been central to marketing, enabling brands to capture attention, inspire emotions and build meaningful connections with consumers. Today, as audiences are exposed to vast amounts of information across multiple channels, creativity can ensure differentiation through engaging and memorable ideas. The rise of artificial intelligence (AI) is reshaping this process and redefining creativity itself. This chapter traces AI's evolution in marketing creativity, from automation to collaboration, highlighting its applications, challenges and future directions, positioning AI as a collaborator in marketing innovation. Overall, this research contributes to the marketing literature by mapping how AI has quickly become an important part of the creative process, and it continues the conversation on the topic by highlighting its implications for theory, practice and future research.

9.1. Introduction

Creativity has always been linked with marketing, as it serves as the driving force that captures attention, sparks emotions and builds lasting connections between brands and consumers. Today more than ever, audiences are constantly being bombarded with an overwhelming amount of messages and content. Furthermore, they have access to vast amounts of information and engage with brands across multiple channels [MOR 25]. Therefore, creativity enables marketers to stand out by delivering fresh, relevant and memorable ideas. Creativity not only drives innovation in campaigns and storytelling but also helps brands to adapt to rapidly changing consumer expectations, cultural trends and technological advancements. Ultimately, creativity transforms marketing from simple

---

Chapter written by Teresa TREVIÑO-BENAVIDES and Flor MORTON.

communication into meaningful engagement, turning products and services into experiences that resonate deeply with people.

While this is relevant, artificial intelligence (AI) comes into play, as its integration into marketing practices is shifting businesses' capabilities to engage with consumers in more targeted, efficient and personalized ways [MOR 25]. Specifically, AI is reshaping the marketing landscape by enhancing and transforming the creative process across multiple areas. From generating personalized content and designing ad visuals to predicting consumer preferences and optimizing campaign strategies, AI tools are increasingly becoming collaborators in the creative process. They can analyze vast amounts of data to uncover insights that inspire novel ideas, suggest innovative formats and even produce text, images or videos that align with brand identity and audience expectations. In doing so, AI not only accelerates creative workflows but also expands the possibilities of imagination, allowing marketers to experiment, adapt and co-create with technology in unimaginable ways.

Particularly, research on the topic has analyzed the evolution of AI in this area of marketing. For example, early applications of AI in the creative process focused only on automating repetitive tasks and analyzing consumer data, enabling marketers to optimize campaigns and personalize content at scale [HUA 20, MUS 20]. At this moment, this represented an important help in becoming more efficient and faster in some routine tasks. Later, as AI matured, it began to augment human creativity, serving as a tool for inspiration, ideation, content generation and campaign design [VAK 20, AME 22, KSH 23, PAG 25].

Taking this into consideration, the objective of this chapter is to examine how AI is transforming marketing creativity by tracing its evolution from automation to co-creation, analyzing its applications and reflecting on its benefits, challenges and future directions for marketers and consumers. It also highlights how AI serves as a partner in augmentation and collaboration, by offering insights, guidance and support to marketers in the creative process. Finally, the chapter addresses the role of personalization and engagement, where AI tools can tailor content to individual preferences, boosting both consumer interaction and brand performance. Together, these mechanisms demonstrate AI's growing role as both a catalyst and collaborator in the evolution of marketing creativity.

Overall, this chapter is structured as follows: first, the literature review will discuss the role of creativity and creative thinking in organizations. Then, it maps and discusses the advances of AI in the creative marketing process by examining recent research since the introduction of AI into organizations and categorizing them in three phases: (a) Mechanical AI; (b) Generative AI (GenAI); and (c) Collaborative AI. Then, the next section will discuss the challenges and limitations of such AI applications in marketing, to finally conclude and reflect on the future directions that companies and marketing teams can expect.

9.2. Literature review

9.2.1. *Creativity and creative thinking in organizations*

There are numerous definitions of creativity, although it is most commonly described as the generation of original ideas that are also useful [BEA 16, PIN 23]. [WAL 19]'s definition of the concept stands out as particularly comprehensive, defining it as "an act arising out of a perception of the environment that acknowledges a certain disequilibrium, resulting in productive activity that challenges patterned thought processes and norms, and gives rise to something new in the form of a physical object or even a mental or an emotional construct" (p. 6). In this definition, four key elements are involved: (1) creativity is an act, not just a result; (2) it is a form of production, not reproduction, since it generates something new rather than repeating the old; (3) it emerges from acknowledging disequilibrium in the environment; and (4) it requires sensitivity to problems or opportunities that others may overlook.

Building on these definitions, it is also important to consider how creativity operates at the cognitive level. Cognition refers to how humans obtain, organize, process and store information [PIN 23]. Creative cognition, therefore, refers to the set of cognitive processes that enable people to generate and evaluate original, useful ideas across different domains [BEA 16]. Creative thinking is a cognitive process that integrates both divergent processes (idea generation) and convergent processes (idea evaluation and refinement) [ZHU 20, PIN 23] and enables the generation of novel and useful ideas. From this perspective, creative thinking has long been recognized not only as a cognitive mechanism but also as one of the essential competencies of the 21st century, crucial for fostering flexibility and enabling individuals and organizations to respond effectively to the opportunities and challenges of an ever-changing environment [RIT 17]. Its pivotal role in driving innovation underscores the pressing need to strengthen creative skills in both educational and organizational settings, and perhaps at no other time has it been so broadly recognized as a virtue across academia, education, the arts and industry as in the present era [GRE 24].

In marketing specifically, creative thinking plays a crucial role in addressing consumer-related challenges, driving innovation in products and services, enhancing brand storytelling, improving customer experience design, developing innovative pricing strategies, creating compelling advertising campaigns and producing engaging digital content, among other marketing activities.

Beyond the individual level, creativity has also been examined through the lens of throughput systems, where organizations are defined by the processes they use to transform resources into outputs or products; in marketing, these processes involve translating market knowledge, audience insights, client objectives and situational constraints into creative concepts [STU 19]. The creative output is not the result of individual effort alone but rather of collaboration, relying on team diversity, specialized expertise and the effective integration of available knowledge [LYN 17].

To further understand how these dynamics unfold, the creative process has been conceptualized through diverse models, each shaped by the level of analysis – whether intrapersonal, organizational or societal – and by the emphasis placed on specific operational dynamics, such as information flow, sequential steps or decision pathways [STU 19]. Among these, sequential or stage-based models are particularly relevant, as they describe creativity as a series of stages or steps that follow a logical and progressive order, making it easier to understand and apply in contexts such as marketing and innovation.

In today's competitive environments, however, creativity alone is not sufficient. As [VEG 15] emphasizes, product innovations are increasingly vulnerable to rapid imitation, which means they often deliver only short-lived competitive advantages. Consequently, the sustained success of a firm does not depend solely on the novelty of individual products, but rather on its capacity to manage and cultivate innovation as a systematic and ongoing process. This perspective underscores the importance of addressing innovation across multiple dimensions of the business model – not only products but also processes, marketing strategies and organizational structures. By doing so, firms can develop the agility required to respond effectively to evolving customer needs.

Unlike product-level innovations, which competitors can replicate with relative ease, a comprehensive innovation capability embedded within the organization is far more difficult to imitate. Such an integrated approach generates a more enduring source of competitive advantage [CHA 18, RAM 21]. For marketing and branding specifically, this implies that sustainable differentiation is less about isolated creative outputs and more about building dynamic systems that consistently enable customer-centric innovation.

In this regard, AI offers a powerful enabler for the type of systematic and multidimensional innovation that organizations require. By transforming creativity into a process that can be structured, analyzed and continuously improved, AI can reduce reliance on sporadic inspiration and provide data-driven support for decision-making. In marketing, AI-driven tools can detect emerging consumer trends, personalize campaigns at scale, optimize content performance in real time and generate new design or storytelling alternatives. This integration of human creativity with machine intelligence has the potential to improve organizational agility and strengthen innovation capabilities that are difficult for competitors to replicate, thereby contributing to more sustainable competitive advantages.

The next section will discuss the trajectory of AI's evolving creative role in marketing, from early marketing automation, personalization, as well as augmentation and collaboration.

9.2.2. *Phase 1 mechanical AI: marketing automation and analytics*

Early applications of AI in the marketing process focused on mechanical AI use for data collection and analytics. Particularly, at this point, AI was being mostly used for the strategic decisions of segmentation, targeting and positioning [HUA 20]. These first applications of AI in marketing were all about automating the mechanical and data-driven tasks. These early systems used AI to collect, track and organize vast amounts of information about markets, competitors, companies and consumers in real time. By automating the repetitive tasks of data sensing, monitoring consumer behavior and analyzing market trends, AI allowed marketers to move from manual data management to more efficient, insight-driven decision-making.

Routine tasks in marketing have traditionally consumed a significant portion of marketers' time and resources. Some of these routine activities include data entry, campaign scheduling, customer segmentation and creating reports. These operational tasks, while essential, take time and often limit the space for creativity. The introduction of AI, particularly through automation and predictive analytics, began to shift this dynamic. By taking over repetitive and time-consuming processes, AI tools allowed marketers to focus their energy on higher-order, creative functions such as ideation, storytelling, brand positioning and consumer experience design.

In this way, the early adoption of AI in marketing marked a transition from purely mechanical marketing operations toward a more balanced model that integrates efficiency with imagination. Routine automation not only enhanced productivity but also expanded creative potential – marketers could use the insights derived from AI-driven data analysis to craft more personalized and emotionally resonant campaigns. Thus, the mechanization of routine marketing tasks through AI did not replace creativity; instead, it amplified it by providing the analytical foundation and time necessary for creative innovation. Although these early uses were operational rather than creative, they laid the groundwork for marketing creativity by freeing up professionals from routine tasks so they could focus on strategic thinking, storytelling and campaign development informed by data intelligence.

9.2.3. *Phase 2 GenAI: creativity enhancement and personalization*

The evolution from mechanical AI to GenAI represents a pivotal shift in how technology supports and enhances marketing creativity. While mechanical AI focuses on automating data-driven and repetitive processes, GenAI focuses on creation. It enables marketers to generate original content, ideas, visuals, copy and even entire campaign ideas. Particularly, GenAI refers to AI systems that can produce content in various formats, including text, images, audio and video. GenAI is designed to generate new data that mimics patterns and structures found in training data. It learns patterns from existing data and generates outputs based on user prompts [BAN 23, BAT 24]. In other words, GenAI generates new content based on past content.

The year 2023 marked an important milestone for GenAI adoption among users and in different contexts [BAT 24]. Recent literature highlights the rapid evolution, broad applications and complex societal, technical and ethical challenges it introduces. In the business and marketing context, GenAI is transforming processes and challenging managers. Particularly, in the creative process, GenAI has been useful in automating content generation, enabling innovation and creating new business models.

Regarding content generation, AI tools can assist by automating certain processes in the creation of marketing materials, including ad copies, images, videos and social media posts. Furthermore, this content creation process is enhanced by the possibility of analyzing large datasets that can produce individual content or "hyper-personalized" campaigns [MOH 25].

A recent study on the topic found that artists and creative managers use AI as an instrument in their creative toolkit. Furthermore, they perceive AI as a catalyst to stimulate new artistic ideas and imaginative thinking and as a tool to deconstruct the creative process [PAG 25]. Research has been focusing on understanding the role of GenAI in creativity. It has been found that GenAI serves as a powerful collaborator that enhances human creativity. The natural concerns about AI replacing creative artists have been dissolving as more integration of AI tools has been incorporated into processes. For example, empirical studies have shown that AI can enhance individual and collaborative creativity, accelerate creative processes and provide new ideas for innovation [EAP 24].

GenAI tool	Main use	Application example in advertising and content creation.
ChatGPT	Generates ad copy, social media posts and text-based content tailored to audiences. Assists in brainstorming and generating ideas for advertising campaigns.	Create ad copy for ads, scripts, headlines, titles, chatbot dialogs, etc.
Copy.ai	Generates marketing copy, including headlines and social media content.	Product descriptions, social media posts, etc.
Jasper	Helps generate ad copy, blog posts and other marketing materials with an SEO focus.	Content for websites and landing pages.
Meta AI	Provides AI tools for generating creative content and enhancing user engagement.	Create interactive content.
DALL-E	Creates high-quality visuals from text descriptions for use in ads.	Create unique and creative product images for ads.
DeepArt.io	Creates stylized images and artwork by applying artistic filters to photos.	Visual ads with artistic effects.
Gemini	Offers advanced image and text generation capabilities for creative content.	Create visually appealing graphics and ad copy for campaigns.
Adobe Firefly	Generates high-quality, customizable images and graphics.	Design ads and visuals.
Playground AI	Generates images and text from prompts, useful for creative projects.	Illustration for posters, mockups, covers, wallpapers and social media copy
Rytr	Assists in writing and generating content for marketing purposes.	Ad copy and social media posts.

Table 9.1. *Generative AI tools summary (source: own elaboration based on [TRE 25])*

Overall, the transformation from mechanical AI to GenAI reflects a move from intelligence that assists human decision-making to intelligence that collaborates in the creative process. GenAI builds on the foundation established by mechanical AI's analytical precision, using vast datasets not only to inform but also to inspire creative output. As a result, creativity in marketing is no longer limited to human intuition alone; it becomes enhanced in a way that human imagination and machine generation intersect to produce more diverse, personalized and impactful brand expressions.

Several GenAI tools can support the creative process. Table 9.1 offers a snapshot of popular GenAI tools, their main uses in the creative process and some application examples in these areas.

9.2.4. *The effects of personalization on consumer experience and behavior*

Considering that GenAI not only can produce content but also focuses on generating content that can appeal to specific audiences, it becomes relevant to talk about personalization and its effects on consumers.

According to [CHA 22], personalization can be understood as the ability to design and create products, services and experiences that align with consumer preferences and needs. It recognizes each customer as a unique individual, with distinct expectations and motivations that require tailored responses that resonate with their tastes, behaviors, motivations and expectations. Beyond a marketing tactic, personalization is regarded as a strategic and technological process that delivers relevant, individualized interactions that enhance the customer experience. These interactions can include product recommendations, customized communication or adaptive content strategies – making personalization a central element to improve customer experience and to build stronger brand–consumer relationships. As an interdisciplinary field, intersecting with business management, computer science, data-driven decision-making, information systems and consumer psychology, personalization enables sophisticated mechanisms to anticipate, interpret and respond to customer expectations, making it a key driver of engagement in the digital era.

Personalization has become one of the key elements of a modern marketing strategy. Its growing significance is reflected not only in improved customer experiences but also in measurable business outcomes: successful personalization initiatives have been shown to increase revenue and enhance marketing efficiency [CHA 22]. Specifically, companies that effectively implement personalization strategies can earn, on average, 40% more revenue from these activities compared to their industry peers [ARO 21]. These figures highlight that personalization is not only a tactical adjustment but also a strategic mechanism that drives business growth. In an era characterized by information overload and competitive parity, the ability to deliver individualized experiences has become essential for building long-term relationships and differentiation. Consumers today increasingly expect brands to understand their unique preferences, motivations and needs – transforming personalization from a marketing advantage into a basic expectation of engagement.

Personalized communication enhances engagement and encourages favorable behavioral intentions, such as repeat purchase or word-of-mouth [NOO 24], improves customer satisfaction, builds stronger emotional bonds and fosters loyalty [OKE 24]. Personalized messages and experiences tend to evoke a sense of relevance and recognition, which strengthens emotional connections with the brand and advertising. However, this relationship is not universally positive. Other studies have suggested that the effectiveness of personalization may diminish when consumers become habituated to such strategies, resulting in desensitization or reduced responsiveness [PFI 20]. Thus, while personalization can drive engagement, satisfaction and loyalty [MOR 24], its long-term success may depend on how well the company balances relevance with novelty and privacy in every interaction. Overexposure to repetitive or overly data-driven personalization can lead to consumer fatigue or even perceptions of manipulation, which can undermine trust in the brand. Therefore, effective personalization strategies should not only leverage data but also maintain ethical principles of transparency, privacy, consumer autonomy and organizational accountability to sustain long-term consumer relationships [MOR 25].

9.2.5. *AI-driven personalization and consumer engagement*

AI has become increasingly central to the design of personalized consumer experiences. By integrating AI, machine learning (ML) and deep learning, firms can obtain actionable insights derived from customer data, both personal and behavioral, which are then used to design and create more relevant experiences. This often requires customer participation – through reviews, purchase histories or social media interactions – that contributes to co-creating value [LIM 22]. Advances in these technologies enable firms to better comprehend and anticipate individual preferences and behaviors, positioning AI as a vital driver of personalization in digital contexts [TEE 25]. Beyond identifying opportunities and understanding customer behaviors, AI-driven personalization aims to create meaningful connections with consumers, extending far beyond traditional marketing approaches [KRI 21].

AI supports engagement marketing by leveraging customer data collected throughout the consumer journey, enabling firms to design more adaptive strategies for value creation. In social media marketing, for instance, AI allows for precise content customization and audience targeting, using data sources such as browsing histories, prior purchases and demographic information [TEE 25]. Recommendation engines illustrate the potential of AI in personalization by analyzing multi-source data to build detailed customer profiles that capture both explicit and implicit preferences [OKE 24]. However, personalization also presents risks if perceived as intrusive or if it compromises consumer privacy, which can lead to negative reactions and eroded trust [ROG 21]. Thus, companies must balance technological efficiency with ethical responsibility.

The integration of AI has transformed the scope and scalability of personalization. AI-powered systems allow firms to analyze vast amounts of customer data, generating insights that support real-time decision-making and highly tailored (or hyper-personalized) interactions with consumers [MOR 24]. According to [CEN 24], these systems identify patterns in consumer behavior, preferences and interactions to deliver customized suggestions that improve the user experience. Their study shows how adaptive interfaces (i.e. dynamic web personalization) driven by ML can dynamically adjust to user behavior in real time, optimizing layout, recommendations and interactions. Their findings indicate that such adaptive systems foster sustained engagement by enhancing accessibility and user satisfaction while reducing the need for manual configuration. In other words, when users interact with an interface that adapts in real time to their behaviors and preferences, they tend to become more engaged and find the experience both more enjoyable and efficient. This adaptability not only strengthens their loyalty to the platform but also motivates them to return, supporting long-term engagement. The authors highlight that the observed link between adaptive design and user satisfaction underscores how essential personalization has become for addressing the varied expectations and needs of users in contemporary digital environments. This not only enhances customer engagement but also improves operational efficiency by optimizing resource allocation. AI-driven personalization thus enables firms to balance mass reach with individual relevance, something that was historically difficult to achieve with traditional marketing techniques.

9.2.6. *Phase 3 collaborative AI: a tool for innovation*

The current evolution in the use of AI within the marketing creativity process moves beyond generative capabilities toward a more collaborative, innovative and human-centered framework. As AI systems advance, their role is shifting from that of a creative assistant to that of a co-creator and innovation partner [DWI 23]. In this emerging stage, AI not only produces content or ideas but also participates in a dynamic exchange of inspiration with human creators. Marketers increasingly rely on AI for real-time ideation, trend forecasting, audience emotion analysis and adaptive storytelling, functions that extend the boundaries of what creativity can achieve [DAV 20, KIE 23].

This new phase of AI use in marketing redefines creativity as an iterative dialogue between human intuition and knowledge and machine intelligence. Rather than replacing creative thinking, AI enhances it by bringing in the capacity to explore endless variations, simulate consumer reactions and propose unconventional solutions that challenge traditional marketing thinking [WIL 18]. The result is a reinvention of the creative process, one that integrates data and innovation into a continuous feedback loop. In this collaborative paradigm, AI becomes an ecosystem for creativity, where marketers, algorithms and audiences co-create value together [TRE 25]. This stage represents not only technological advancement but also a cultural transformation in how creativity is conceptualized, produced and experienced in marketing.

Figure 9.1 shows the evolutionary trajectory of AI within the marketing creativity process.

Moving from mechanical AI to GenAI, and ultimately toward collaborative AI. In the first phase, mechanical AI represents the automation of routine and repetitive marketing tasks, such as data collection, monitoring consumer behavior and performance tracking, allowing marketers and creatives to focus on more strategic functions. In the second phase, GenAI, expands AI's role by enabling the creation of original content, enhancing creative output and efficiency. In this phase, personalization becomes relevant as the content can be tailored for specific audiences. Finally, the collaborative AI phase signifies a paradigm shift toward the relationship between humans and AI, focusing on a deep co-creation process, where technology acts as a partner in ideation, innovation and strategic decision-making.

This progression underscores how AI has evolved from being a tool for automation to becoming a creative and innovative collaborator in marketing practices.

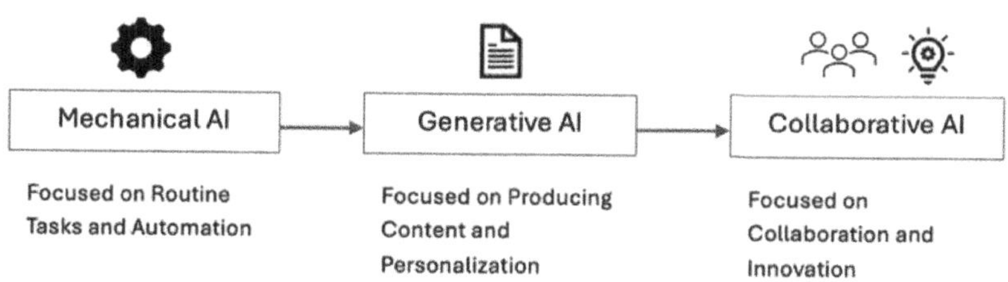

Figure 9.1. *AI's evolution in marketing creativity (source: own elaboration)*

9.3. Challenges and limitations of AI in marketing

While AI's evolution from mechanical to generative and collaborative phases has revolutionized marketing creativity, this transformation also introduces a series of ethical, strategic and cognitive challenges. Each stage of AI's integration presents its own limitations that affect how creativity is executed.

In the mechanical phase, the main issue was the overreliance on automation and data, which risked reducing marketing to a purely analytical exercise. Although automation enhanced efficiency, it often neglected the emotional and cultural dimensions that make creative communication resonate with audiences [DAV 20].

As the field advanced into the generative phase, new challenges emerged around authenticity, originality and intellectual property. GenAI's ability to produce vast amounts of creative content raised concerns about the dilution of brand identity and the loss of human distinctiveness [KIE 23]. The boundary between inspiration and imitation has become increasingly blurred, questioning who truly owns AI-generated ideas and whether they can maintain the emotional depth that human creativity provides [DWI 23]. Moreover, overdependence on AI-generated material may foster creative homogenization, where algorithms trained on similar data produce predictable, culturally narrow or biased outputs. For example, regarding the quality of content produced by AI, a recent study evaluated how GenAI enhances individual creativity but, at the same time, reduces the diversity of novel content. Particularly, it found that content produced by AI was better written and more creative but more similar to each other than content produced by humans alone. This suggests that using AI in the creative process may increase the risk of losing novelty [DOS 23].

Next, in the emerging collaborative phase, the challenges shift toward integration and ethics. While AI–human collaboration offers creative possibilities, it also demands a redefinition of creative authorship, accountability and trust [WIL 18]. Marketers must learn to understand the fine balance between human judgment and AI suggestions, ensuring that creative strategies remain aligned with brand values and societal norms. Popular ethical dilemmas, which include the manipulation of consumer emotions, data privacy risks and the use of synthetic media, may require new frameworks for responsible AI use [TRE 25].

Finally, a significant limitation lies in the human capability gap. The successful integration of AI in creativity depends on marketers' ability to interpret, supervise and challenge machine outputs. Without sufficient digital literacy and creative confidence, human collaborators risk becoming passive users of AI ideas rather than active co-creators [DWI 23]. Thus, while AI continues to expand the creative frontier, its long-term success in marketing depends on nurturing human imagination, critical thinking and ethical awareness alongside technological innovation.

9.4. Future directions of AI in marketing and creativity

The convergence of AI, personalization and creativity is transforming marketing into a more adaptive, human-centered discipline. AI's capacity to analyze vast amounts of data enables firms to design hyper-personalized experiences that evolve in real time, fostering relevance, satisfaction and loyalty. Beyond efficiency, personalization has become a strategic capability that integrates technological intelligence with emotional resonance, emphasizing transparency, privacy and value co-creation through consumer participation. At the same time, creativity remains essential, as AI augments rather than replaces human imagination, enabling new forms of experimentation and adaptive design [MOR 25] that align data-driven precision with authentic connection.

Future directions in AI and creativity point toward a deeper integration between computational intelligence and human imagination, where technology not only automates tasks but also actively collaborates in the creative process. As generative models and other systems evolve, AI will increasingly support ideation, design and storytelling that respond dynamically to emotional and contextual cues. Rather than replacing human creativity, AI can function as a cognitive partner, enhancing divergent thinking, accelerating prototyping and enabling more inclusive participation in creative production. Ethical considerations, however, will remain central, requiring transparency, authorship recognition and accountability in AI-assisted creation. Ultimately, the future of AI and creativity depends on fostering close collaboration between human imagination and machine intelligence, where new ideas emerge from their interaction rather than their opposition. Together, AI and creativity will redefine personalization as more than a marketing technique, becoming a dynamic process of co-creation that integrates analytical insight with human meaning, shaping the way consumers experience and engage with brands.

9.5. Implications and future research

This chapter intend to discuss the evolution of AI's role in marketing creativity. Particularly, the research dived into marketing literature to understand how AI use has been evolving from mechanical automation to more collaborative and innovative. This chapter continued the conversation on the topic and positions AI not as a replacement for human creativity but as a transformative partner reshaping the boundaries of marketing creativity.

As for the implications, for practitioners, the findings highlight the importance of blending new capabilities with analytical reasoning with human intuition and creativity. Marketers must learn not only to use AI tools but also to guide, question and humanize AI outputs. Organizations must adopt processes, employee education and the overall culture to foster AI-augmented creative ecosystems that consider managers' insights, technological innovation and ethical guidelines.

To continue advancing on the theoretical implications, future research can assess the influence of AI on marketing effectiveness and evaluate the impact of AI on decision-making in a creative marketing context [PAG 25]. Few studies have actually evaluated the perceptions and effectiveness of marketing campaigns created by humans versus those created by AI. Particularly, it can be addressed how audiences interpret and respond to content created by AI compared to content created by humans [LON 19, LUO 23]. Research on the topic can help advance conversations about expectations and perceptions of consumers when exposed to such content. Furthermore, it becomes interesting to academics to develop conceptual frameworks that explain how creativity is being redefined when the content results from human–AI collaboration [CRO 23, FLO 24].

Finally, considering the third phase of AI's evolution in marketing creativity, and understanding that to achieve a real collaboration between humans and AI, it is necessary to understand the implications for education and talent development. Thus, future research can focus on understanding new skills, capabilities and styles in marketing professionals that will be necessary to develop in creative industries enhanced with AI [KAP 22].

9.6. References

[AME 22] AMEEN N., SHARMA G., TARBA S. et al., "Toward advancing theory on creativity in marketing and artificial intelligence", *Psychology & Marketing*, vol. 39, no. 9, pp. 1802–1825, doi: 10.1002/mar.21699, 2022.

[ARO 21] ARORA N., ENSSLEN D., FIEDLER L. et al., The value of getting personalization right – or wrong – is multiplying, Report, McKinsey & Company, available at: https://www.mckinsey.com/capabilities/growth-marketing-and-sales/our-insights/the-value-of-getting-personalization-right-or-wrong-is-multiplying, 2021.

[BAN 23] BANH L., STROBEL G., "Generative artificial intelligence", *Electronic Markets*, vol. 33, pp. 1–17, doi: 10.1007/s12525-023-00680-1, 2023.

[BAT 24] BATCHU C., SATYA V., "Generative AI: Evolution and its future", *International Journal for Multidisciplinary Research*, vol. 6, no. 1, doi: 10.36948/ijfmr.2024.v06i01.12046, 2024.

[BEA 16] BEATY R.E., BENEDEK M., SILVIA P.J. et al., "Creative cognition and brain network dynamics", *Trends in Cognitive Sciences*, vol. 20, no. 2, pp. 87–95, 2016.

[CEN 24] CEN Z., ZHAO Y., "Enhancing user engagement through adaptive interfaces: A study on real-time personalization in web applications", *Journal of Economic Theory and Business Management*, vol. 1, no. 6, pp. 1–7, 2024.

[CHA 18] CHATZOGLOU P., CHATZOUDES D., "The role of innovation in building competitive advantages: An empirical investigation", *European Journal of Innovation Management*, vol. 21, pp. 44–69, doi: 10.1108/EJIM-02-2017-0015, 2018.

[CHA 22] CHANDRA S., VERMA S., LIM W.M. et al., "Personalization in personalized marketing: Trends and ways forward", *Psychology & Marketing*, vol. 39, no. 8, pp. 1529–1562, 2022.

[CRO 23] CROOKS R., GHOSH A., TADDY M., "Generative AI and the creative economy", *Journal of Marketing*, vol. 87, no. 5, pp. 1–15, 2023.

[DAV 20] DAVENPORT T., GUHA A., GREWAL D. et al., "How artificial intelligence will change the future of marketing", *Journal of the Academy of Marketing Science*, vol. 48, no. 1, pp. 24–42, 2020.

[DOS 23] DOSHI A., HAUSER O., "Generative AI enhances individual creativity but reduces the collective diversity of novel content", *Science Advances*, vol. 10, doi: 10.1126/sciadv.adn5290, 2023.

[DWI 23] DWIVEDI Y.K., HUGHES L., BAABDULLAH A.M. et al., "Artificial Intelligence (AI): Multidisciplinary perspectives on emerging challenges, opportunities, and agenda for research, practice, and policy", *International Journal of Information Management*, vol. 71, p. 102642, 2023.

[EAP 24] EAPEN T., FINKENSTADT D., FOLK J. et al., "How generative AI can augment human creativity", *SSRN Electronic Journal*, doi: 10.2139/ssrn.4759930, 2024.

[FLO 24] FLORIDA R., ADLER P., MELLANDER C., "Creative collaboration in the age of AI", *Creativity Research Journal*, vol. 36, no. 1, pp. 45–61, 2024.

[GRE 24] GREEN A.E., BEATY R.E., KENETT Y.N. et al., "The process definition of creativity", *Creativity Research Journal*, vol. 36, no. 3, pp. 544–572, 2024.

[HUA 20] HUANG M., RUST R., "A strategic framework for artificial intelligence in marketing", *Journal of the Academy of Marketing Science*, vol. 49, pp. 30–50, doi: 10.1007/s11747-020-00749-9, 2020.

[KAP 22] KAPITAN S., SILVERA D.H., HABIBI M.R., "Human – AI collaboration: Reimagining creativity and ethics in marketing", *Journal of Business Research*, vol. 145, pp. 658–668, 2022.

[KIE 23] KIETZMANN J., PITT L., PASCHEN J. et al., "Artificial intelligence in advertising: How marketers can leverage generative AI for creative advantage", *Journal of Advertising Research*, vol. 63, no. 2, pp. 103–118, 2023.

[KRI 21] KRISHEN A.S., DWIVEDI Y.K., BINDU N. et al., "A broad overview of interactive digital marketing: A bibliometric network analysis", *Journal of Business Research*, vol. 131, pp. 183–195, 2021.

[KSH 23] KSHETRI N., DWIVEDI Y., DAVENPORT T. et al., "Generative artificial intelligence in marketing: Applications, opportunities, challenges, and research agenda", *International Journal of Information Management*, vol. 75, p. 102716, 2023.

[LIM 22] LIM W.M., RASUL T., KUMAR S. et al., "Past, present, and future of customer engagement", *Journal of Business Research*, vol. 140, pp. 439–458, 2022.

[LON 19] LONGONI C., BONEZZI A., MOREWEDGE C.K., "Resistance to algorithms and AI in decision-making", *Journal of Marketing Research*, vol. 56, no. 4, pp. 615–635, 2019.

[LUO 23] LUO X., TONG S., FANG Z. et al., "Frontiers in artificial intelligence and marketing: Past, present, and future", *Journal of Marketing*, vol. 87, no. 1, pp. 84–109, 2023.

[LYN 17] LYNCH J., WEST D., "Agency creativity: Teams and performance: A conceptual model links agency teams' knowledge utilization, agency creativity, and performance", *Journal of Advertising Research*, vol. 57, no. 1, pp. 67–81, 2017.

[MOH 25] MOHARAM M.M.R., TAWALBEH A., "The role of gen AI in enhancing creativity and efficiency in content marketing creation: Scoping review and future insights", *International Journal of Innovative Research and Scientific Studies*, vol. 8, no. 1, pp. 2804–2814, 2025.

[MOR 24] MORTON F., TREVIÑO T., GONZÁLEZ-TREVIÑO E., "Taking customer-centricity to new heights: Exploring the intersection of AI, hyper-personalization, and customer-centricity in organizations", in MACHADO C., DAVIM J.P. (eds), *Smart Engineering Management*, Springer, Cham, 2024.

[MOR 25] MORTON F., TREVIÑO-BENAVIDES T., "The role of AI in planning, executing, and evaluating IMC", in MACHADO C., DAVIM J.P. (eds), *Challenges and Opportunities in the Artificial Intelligence Era*, Springer, Cham, 2025.

[MUS 20] MUSTAK M., SALMINEN J., PLÉ L. et al., "Artificial intelligence in marketing: Topic modeling, scientometric analysis, and research agenda", *Journal of Business Research*, doi: 10.1016/j.jbusres.2020.10.044, 2020.

[NOO 24] NOOR U., MANSOOR M., SHAMIM A., "Customers create customers!: Assessing the role of perceived personalization, online advertising engagement and online users' modes in generating positive e-WOM", *Asia-Pacific Journal of Business Administration*, vol. 16, no. 2, pp. 392–409, 2024.

[OKE 24] OKEKE N.I., ALABI O.A., IGWE A.N. et al., "AI-driven personalization framework for SMEs: Revolutionizing customer engagement and retention", *World Journal of Advanced Research and Reviews*, vol. 24, no. 1, pp. 2019–2035, 2024.

[PAG 25] PAGANI M., WIND Y., "Unlocking marketing creativity using artificial intelligence", *Journal of Interactive Marketing*, vol. 60, pp. 1–24, doi: 10.1177/10949968241265855, 2025.

[PFI 20] PFIFFELMANN J., DENS N., SOULEZ S., "Personalized advertisements with integration of names and photographs: An eye-tracking experiment", *Journal of Business Research*, vol. 111, pp. 196–207, 2020.

[PIN 23] PINKOW F., "Creative cognition: A multidisciplinary and integrative framework of creative thinking", *Creativity and Innovation Management*, vol. 32, no. 3, pp. 472–492, 2023.

[RAM 21] RAMPA R., AGOGUÉ M., "Developing radical innovation capabilities: Exploring the effects of training employees for creativity and innovation", *Creativity and Innovation Management*, vol. 30, pp. 211–227, 2021.

[RIT 17] RITTER S., MOSTERT N., "Enhancement of creative thinking skills using a cognitive-based creativity training", *Journal of Cognitive Enhancement*, vol. 1, pp. 243–253, 2017.

[ROG 21] ROGGEVEEN A.L., GREWAL D., KARSBERG J. et al., "Forging meaningful consumer-brand relationships through creative merchandise offerings and innovative merchandising strategies", *Journal of Retailing*, vol. 97, no. 1, pp. 81–98, 2021.

[STU 19] STUHLFAUT M.W., WINDELS K., "Altered states: The effects of media and technology on the creative process in advertising agencies", *Journal of Marketing Communications*, vol. 25, no. 1, pp. 1–27, 2019.

[TEE 25] TEEPAPAL T., "AI-driven personalization: Unraveling consumer perceptions in social media engagement", *Computers in Human Behavior*, vol. 165, pp. 1–9, 2025.

[TRE 25] TREVIÑO-BENAVIDES T., MORTON F., "AI revolutionizing marketing: Applications and other considerations", in MACHADO C., DAVIM J.P. (eds), *Challenges and Opportunities in the Artificial Intelligence Era*, Springer, Cham, 2025.

[TUR 17] TURNBULL S., WHEELER C., "The advertising creative process: A study of UK agencies", *Journal of Marketing Communications*, vol. 23, no. 2, pp. 176–194, 2017.

[VAK 20] VAKRATSAS D., WANG X., "Artificial intelligence in advertising creativity", *Journal of Advertising*, vol. 50, pp. 39–51, doi: 10.1080/00913367.2020.1843090, 2020.

[VEG 15] VEGA-JURADO J., JULIAO-ESPARRAGOZA D., PATERNINA-ARBOLEDA C.D. et al., "Integrating technology, management and marketing innovation through open innovation models", *Journal of Technology Management & Innovation*, vol. 10, no. 4, pp. 85–90, 2015.

[WAL 19] WALIA C., "A dynamic definition of creativity", *Creativity Research Journal*, vol. 31, no. 3, pp. 237–247, 2019.

[WIL 18] WILSON H.J., DAUGHERTY P.R., "Collaborative intelligence: Humans and AI are joining forces", *Harvard Business Review*, vol. 96, no. 4, pp. 114–123, 2018.

[ZHU 20] ZHUANG K., YANG W., LI Y. et al., "Connectome-based evidence for creative thinking as an emergent property of ordinary cognitive operations", *NeuroImage*, vol. 227, pp. 1–15, 2020.

10

Unveiling Management Research's Thematic Evolution: An Unsupervised Machine Learning – Latent Dirichlet Allocation Perspective

In the dynamic realm of organizational practices and strategies, navigating complexity demands robust methodologies unveiling intricate themes within management research. This study employs advanced topic modeling, particularly Latent Dirichlet Allocation (LDA), to address: "What thematic patterns emerge in management research?" Unearthing latent themes holds scholarly and pragmatic value. Scholars and practitioners seek insights into trends shaping decisions and strategies. This inquiry delves into concealed research architecture, guiding them as they navigate management research. While existing wisdom contributes invaluable frameworks and insights, understanding themes across diverse articles remains challenging. This study leverages LDA's power, employing unsupervised learning for dynamic thematic landscapes, revealing insights into paramount themes. At its core, this study triggers scholarly transformation. Harnessing LDA, it transcends content analysis, heralding data-driven insights into latent themes and their arcs. The narrative is amplified through post-LDA visualization, illuminating enigmatic theme substrata. Innovation continues with multiple correspondence analysis (MCA), harmonizing keywords and topics. This strengthens thematic connections, adding quantitative support to qualitative exploration. By analyzing influential authors and papers through citations, a constellation emerges, shedding light on contributors and milestones sculpting management's evolution.

Chapter written by Qeis KAMRAN.

For a color version of all the figures in this chapter see, www.iste.co.uk/machado/Artificial.zip.

NOTE.– Given the scale and complexity of the datasets underlying Figures 10.9 to 10.15, the complete high-resolution and interactive representations of these figures are made available through a linked GitHub repository, accessible via the QR code at the end of this chapter. To further ensure transparency and reproducibility, these full-resolution visualizations are provided in an interactive HTML environment that allows readers to explore the complete figure set beyond the limitations of print reproduction.

In summary, this study establishes a comprehensive framework for topic modeling in management research. Empowering scholars and practitioners, it fosters the traversal of management literature, shaping scholarship and strategies, and enhancing innovation in theory and practice. These insights navigate the intricate tapestry of management research, providing a roadmap for understanding themes, trends and their influence. An additional field of application pertaining to the design for a serial entrepreneurship in cyber-physical realities – deciphering visionary management models for marketing – has been provided to underpin the diagnostic power of the developed methodology of meta-level research.

10.1. Introduction

Management research plays a pivotal role in shaping organizational practices and strategies across various industries. The continuous evolution of this field necessitates robust methodologies that can unveil key themes and trends in the vast corpus of management literature. In this study, the author employs advanced topic modeling techniques, specifically Latent Dirichlet Allocation (LDA), to address the pressing research question: "What are the dominant thematic patterns and how have they evolved in management research over time?"

Understanding the underlying themes in management research is crucial both in theory and practice. Academics and practitioners seek to stay abreast of emerging trends and focus on areas that influence decision-making and business strategies. By unraveling the latent structures within the corpus, the author aims to provide valuable insights that can aid scholars, practitioners and journal editors in navigating the dynamic landscape of management research.

(1) What do we know?

Prior research in management has provided valuable theoretical perspectives and empirical findings, which have shaped our understanding of various subfields. However, the complexity and volume of management literature have made it challenging to comprehensively grasp the evolving themes and trends. While existing studies have shed light on specific topics, a major unaddressed puzzle remains: How do we holistically capture the thematic patterns and shifts across a vast corpus of management research articles?

This study addresses this crucial knowledge gap by employing LDA, a powerful topic modeling algorithm, to uncover key themes and their temporal evolution. By delving into the unsupervised learning approach, we provide a novel and comprehensive perspective on the dynamics of management research themes. This work holds great significance, as it facilitates a deeper understanding of the overarching themes that influence scholarly discourse and practical applications.

(2) What will we learn?

Through this study, we hope to fundamentally change and advance scholars' understanding of management research. By applying LDA, it goes beyond traditional content analysis and provides a data-driven approach that enables the identification of latent themes and their transitions over time. The post-LDA processing and visualization techniques enhance the interpretability of results, fostering a deeper understanding of the underlying thematic structures.

Furthermore, our application of multiple correspondence analysis (MCA) offers a quantitative validation of the alignment between article keywords and topic assignments. This innovative approach contributes to a more robust assessment of the thematic associations.

By identifying the most influential authors and papers based on citation counts, the research sheds light on key contributors and landmark publications that have significantly impacted the development of the management field.

Overall, our study offers a comprehensive and systematic framework for topic modeling and analysis in management research, empowering scholars and practitioners to explore, interpret and navigate the ever-changing landscape of management literature. The insights gained from this research have the potential to shape future scholarly work, influence strategic decisions in organizations and foster innovation in the management field.

10.2. Method

The method proposed within this research lays out numerous essential aspects and necessary dimensions to implement itself as an important contribution for future researchers, who either wish to cope with a large amount of data or would like to deliver impactful and solidified research contributions within their research field. The aim is to avoid bias in conducting research. Therefore, all keywords used in research articles have been put through an unsupervised machine learning algorithm to establish the historical and in-depth content analysis. Typically, conceptual articles cannot entirely renounce themselves from conceptual bias, and therefore, the analysis through a machine learning algorithm and content analysis delivers the desired results for researchers.

For this research, the total amount of textual data of 21 journals has been analyzed with the help of a topic extraction algorithm to compose and provide an objective representation of the historical narrative. Therefore, the scholarly analyzed "*lay of the land*" in management has been combined with machine learning algorithm (MLA) techniques and tools to allow for more precision and objectivity by establishing unbiased labeling of the "*topic-keyword relations*". To be precise, an "LDA" has been used [BLE 03]. It is a statistical model and collection of documents [BLE 12] and raises the assumption that each topic is associated with multiple terms and reversed, terms can also be associated with more than one topic but have a different impact on the overall topic distribution. According to Tufts [TUF 17], LDA can be interpreted as follows:

– Topic structures in a document are latent, implying hidden structures within the text.

– The Dirichlet distribution is assumed to determine the mixture proportions of the topics in the documents and the words in each topic.

– Assignment of words to a specific topic [TUF 17].

Based on the developments in machine learning capacity, this methodology provides a solid and robust foundation for research in which large amounts of textual data need to be analyzed to create a more holistic grounded design theory. Therefore, LDA illustrates a probabilistic generative model, which takes the assumption that each document is a distribution over topics and each topic is a distribution over words [ARU 10]. A dynamic approach is being delivered to emphasize and highlight the evaluation of the topics over time.

A content analysis of all papers published in the selected journals can first be performed, however editorial notes, errata or commentaries should be excluded to thoroughly examine the content of management literature, trace its evolution and identify main streams and subfields. A subsequent analysis examines the relation of topics over varying time horizons. It reflects the evolution of the respective field and provides journal editors, reviewers and authors with an interpretation of the direction of the respective journal [FUR 08]. This analysis is being used for an objective, systematic and quantitative consideration of all selected articles, while at the same time allowing for an interpretation of the shifting priorities of all parties involved, namely editors, reviewers and authors, who shaped the evolution of the management field.

As a first step, the article universe has to be selected. Meta-information such as the title, author, keywords and abstracts of every journal article with the highest impact factor should serve as a base. Note that only a large number of total articles (in our opinion $n >> 1{,}000$ should serve as a threshold) justify the use of MLA.

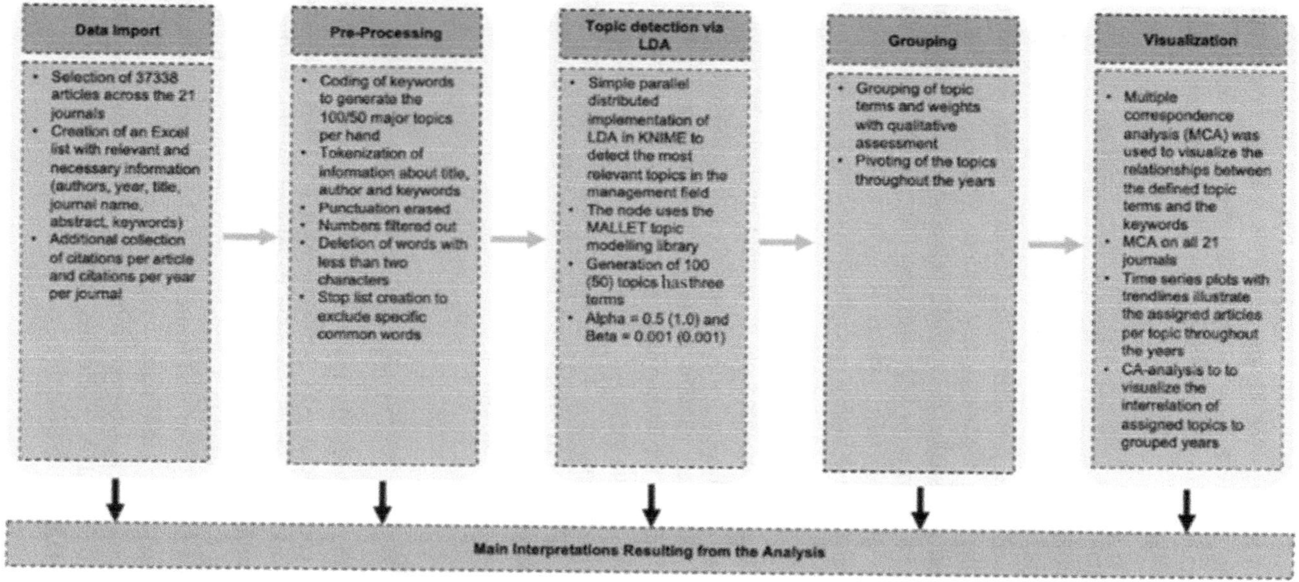

Figure 10.1. *Exemplary workflow of an LDA topic extraction with subsequent content analysis*

In Figure 10.1, this research shows an exemplary research project reading out $n = 37{,}331$ management articles over 21 journals. This research investigated two different topic models, the first with $K = 50$ topics and the second with $K = 100$ topics that were automatically extracted by the LDA algorithm. For this, proper pre-processing has to be installed.

Further hyper-parameters such as α and β have to be estimated a priori. α is an indicator for the document-topic distribution, that is assumed to follow a special probability distribution, that is called *Dirichlet distribution*. The higher the parameter α the more topics are assumed to be prevalent in the research articles. Accordingly, β measures the a priori belief of a topic-term distribution, which is also assumed to follow a Dirichlet distribution. Usually, β is small, i.e. close to zero, indicating that there are only a few numbers of terms (or keywords) that fully identify a research topic, i.e. the topic content is quite specific, which is usually the case for research articles since academics tend to use highly specialized vocabulary in their respective fields.

In a second phase, the author conducts a content analysis by pivoting the topic-labeled articles over different year horizons (usually five years) and by counting them. This research has identified patterns of different topics being prominently discussed in certain years. Finally, the author highlights the results in a bubble diagram showing year group bubbles versus extracted topics. We can see topics treated more frequently in the past and new emerging topics such as digitalization and corporate social responsibility (CSR). Moreover, evergreen topics, being discussed almost every decade, such as pricing, performance and other operation research topics, are found in the center of the diagram (see Figure 10.10).

10.2.1. *Pre-processing*

For the execution of the analyses, the machine learning software "KNIME" can be used. KNIME is a commercial API software with an MLA drag-and-drop solution. The KNIME workflow, which has been developed for this methodology, consists of 47 nodes in total, shown in Figure 10.2. A node is a pre-implemented snippet of algorithms that can be used in a drag-and-drop environment for building the overall analysis workflow. These nodes are divided into nine "meta-nodes" (combining single nodes) after the data has been sent through the LDA. Nevertheless, before generating a reliable and robust output, the previously collected data must be optimized for the LDA. Therefore, eight nodes are necessary to reach the ideal dataset for the LDA.

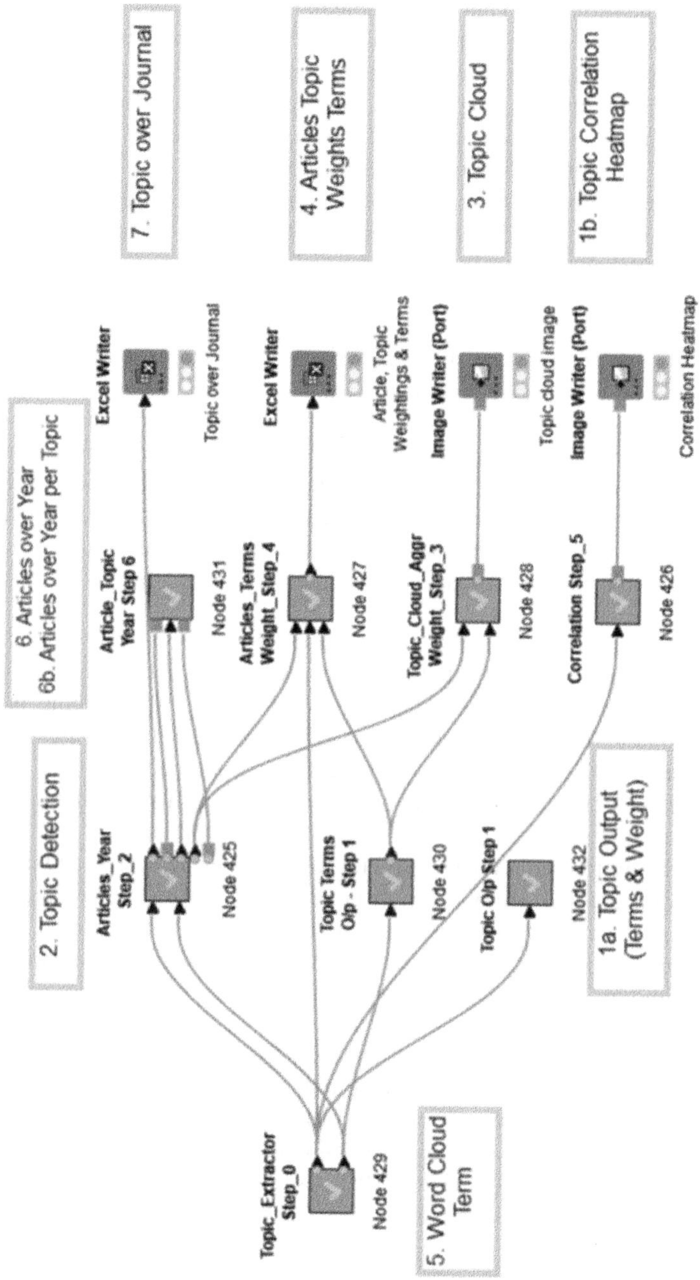

Figure 10.2. *Overall KNIME workflow subsumed in meta-nodes*

The first node is the "*Excel Reader*", which reads Excel files (either single or multiple files at the same time) but is only able to read one sheet per time. The Excel types that are supported are string, number, Boolean, formulas, date and time. Pictures and diagrams cannot be read. The data is then read in and converted into KNIME types: string, integer, long, double, Boolean, local date, local time and local date and time. The node will scan the input file and determine the number and types of the columns. The output then represents a table with the auto-guessed structure and KNIME types. Before starting the node, it is important to notice that the presets of the node will only cover 15,000 rows. In the advanced settings of the node, it is possible to set this number to the desired number (the total number of articles) [HUB n.d. a].

The second node converts the specified strings into documents. A document will be created for each row and then attached to that specific row. The strings of the columns will be used as title, authors and full text, while additionally the defined category, source type and data will be set. Since this research aims to determine the correlation of articles and keywords to a specific topic, the text section of the node will be assigned to the keywords, which are stated in the input document. The keywords can be retrieved from the databases *Web of Science* and through the websites of the publishers of the respective journals [HUB n.d. b].

The information about the title, author and keywords is transformed into a document data type and then adjusted by using a word tokenizer (*OpenNLP English WordTokenizer*). The so-called tokenization is the process of chopping the given sentence into smaller parts, the tokens, which is used in tasks such as spell-checking, processing searches, identifying parts of speech, sentence detection and document classification [TUT 21]. Nodes three to six remove all punctuation characters in the input document, filters out all digits, including decimal separators "," or "." and "+" or "-", and terms in the input document with less than *N* characters, and convert all terms in the input document into lower- or upper-case letters, respectively. All terms with less than three characters in the input document have to be deleted, all letters have to be turned into lower case letters, and the raw text has to be tokenized based on a set of delimiters, for example, whitespaces and punctuation.

The next steps include the integration of a built-in "stop list" into KNIME, containing specific words that are very common in the English language, and a manual one with additional very common words in the field of study. Those words are excluded in the analysis due to the likelihood of the distortion of the results given their weight and impact. Figure 10.3 visualizes the first nine nodes in the KNIME workflow used for this research.

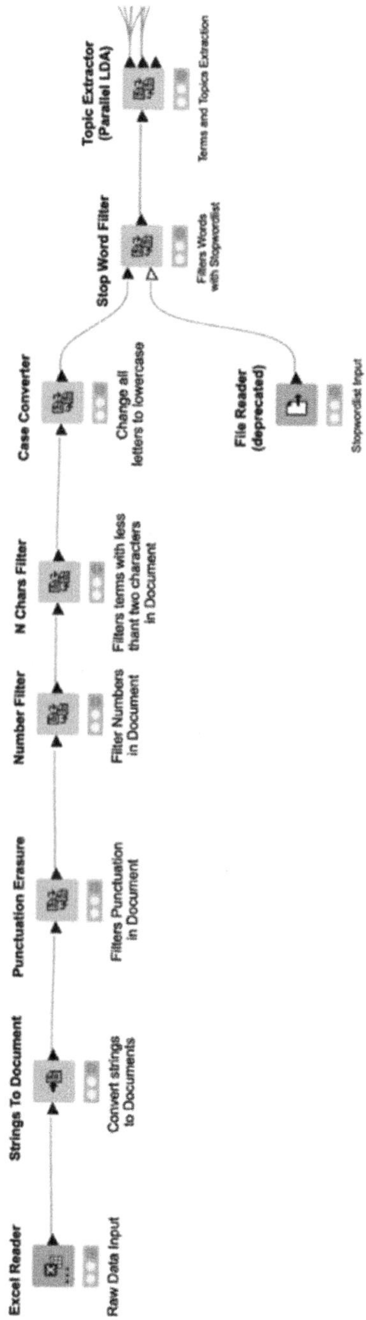

Figure 10.3. *Pre-processing steps in KNIME, visualization of the first nine nodes including the LDA*

Once all eight of these nodes have been successfully performed, the LDA can be configured and started.

10.2.2. *Topic detection via LDA*

To analyze and detect the most prominent topics in the field of study during the chosen time span, a simple parallel distributed implementation of LDA in accordance with Newman et al. [NEW 09] is used with the addition of the sparse LDA sampling scheme and data structure according to Yao et al. [YAO 09].[1] This node is the second node in the workflow and represents the core of the model. Therefore, LDA needs an a priori defined number of topics *(K),* the number of words or terms (*T*) as commonly used in LDA literature, associated with the topic, as well as additional distribution parameters alpha (α) and beta (β) as described above. This serves the purpose of strongly favoring sparse word distributions. Again, a high alpha is used for documents that cover many topics and therefore have a lower impact of topic sparsity. A low alpha is being used in cases where only a few topics are covered, but with a higher impact of topic sparsity. Usually, alpha is set to a fraction of the number of *K* [KON 18]. Figure 10.4 shows the topic sparsity.

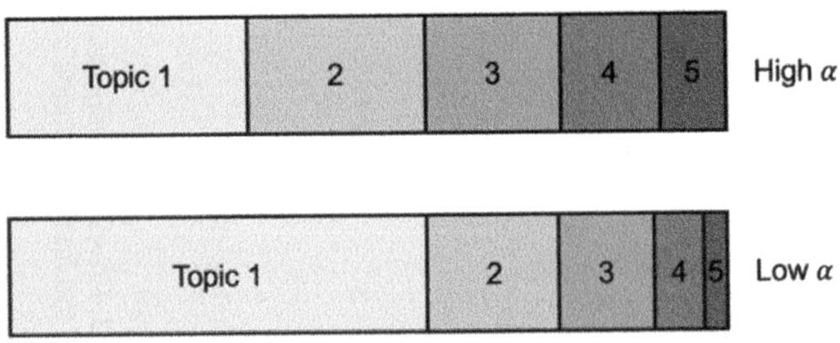

Figure 10.4. *Topic sparsity for high and low alpha*

If a high beta is being used, each topic consists of many words, having a low impact of word sparsity, and therefore, the output will cover more general topics. However, when a low beta is being used, each topic consists of a few words, hence a higher impact of word sparsity, thus more specified topics. Beta is being used for the granularity of term-topic distribution, indicating that a high beta allows for fewer more general topics, while a low beta allows for more specific topics [KON 18]. Figure 10.5 shows the word sparsity.

1 From the authors' experience, increasing the number of terms per topic leads to a blurred interpretations of the overall topic detection and is inevitably linked to an increased effort in preprocessing (e.g. customization of the stop word list). The intention was to reduce the preprocessing amount to as few steps as possible.

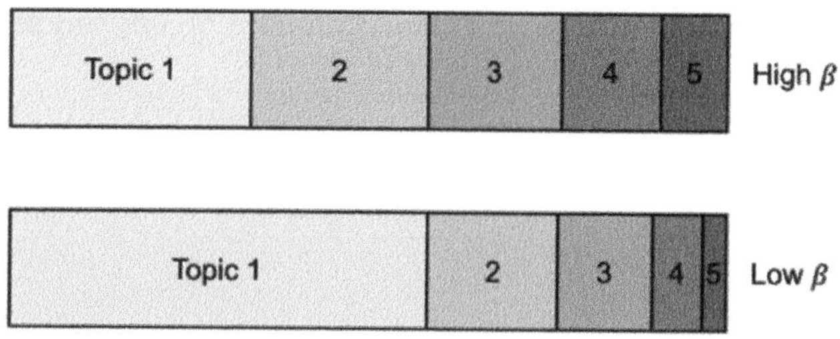

Figure 10.5. *Word sparsity for high and low beta*

Nevertheless, to determine the right alpha and beta values, three sensitivity analyses, for coarse, fine and medium granularity, can be performed as a preliminary analysis. Hereby, it has been decided to use the example of the coarse granularity.

Figure 10.6 shows a foregoing sensitivity analysis for a coarse granularity of topics, i.e. by using higher alpha and beta values and by evaluating various "goodness of fit" measures for LDA topic modeling. For example, the "coherence", which was pre-dominantly used in our studies, measures how well the extracted keywords characterize each topic. The higher the coherence, the better the fit. There are also other evaluation measures, such as the so-called "perplexity," that are optimal for our research if they are minimized, i.e. the lower the perplexity scores, the better the fit. However, recent studies show that perplexity is often counter-intuitive to human language [CHA 09]. Unfortunately, in real-life topic extraction projects not all evaluation measures tend to equally point in the same evaluation direction.

As we can see in Figure 10.6 once again, the perplexity score (to be minimized) would speak for a very small number of topics to be extracted, whereas the measure proposed by Cao et al. [CAO 09] indicates a very high number of topics. Arun et al. [ARU 10] even has a global minimum for $K = 50$ topics. Similarly, for the coherence measures (to be maximized). The standard (textbook method) coherence measure, as pre-implemented in the Python package *Gensim*, speaks for a higher number of topics, whereas the coherence measure proposed by Mimno et al. [MIM 11] indicates a smaller number of topics. As already indicated, in our research projects realized so far, usually the Genism coherence correlates to our human judgment when we qualitatively evaluate the proposed research topics with respect to their associated keywords. For other research projects, it may be valuable to also keep the other evaluation measures in mind. Therefore, this research would like to highlight every single evaluation measure in due brevity.

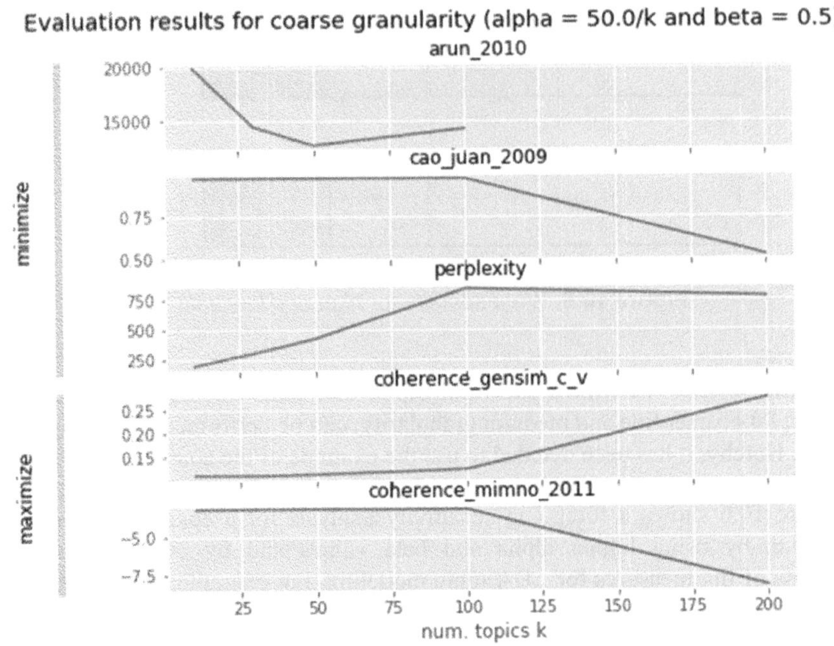

Figure 10.6. *Evaluation results for coarse granularity*

Arun et al. [ARU 10] present empirical results indicating that the "graph dips down and hits a low" for the "right" number of topics and increases as the number of topics also increases. Therefore, the right number of topics is any number in a small range giving the best accuracy out of the respective dataset. In this case, in accordance with Arun et al. [ARU 10], 50 topics seem to be the right number of topics. Nevertheless, the value is still low even for 100 topics.[2]

Cao et al. [CAO 09] propose that the LDA model performs best when the average cosine distance of topics reaches a minimum; hence, this graph should be minimized. They find the best number of topics based on the topic density. In this research, the graph starts minimizing once it has reached 100 topics. Therefore, this method would suggest using more than 100 topics, reaching its lowest point with approximately 200 topics.

2 The Arun et al. measure was not defined for a topic number higher than 100 due to internal algorithmic constraints.

The third measure deals with perplexity. Perplexity is a commonly used indicator, where a lower perplexity indicates a better prediction [BLE 03, JAC 16]. For this case, the graph shows how the perplexity increases as the number of topics increases. Once the graph has reached 100 topics, it starts decreasing. Therefore, the perplexity method also suggests using as many topics as possible or as few topics as possible.

The fourth measure deals with coherence. Röder et al. [RÖD 15] imply that coherence measures have gained importance in text mining and unsupervised learning methods, such as topic modeling (proposed in this research), since these methods give no guarantees on the interpretability of the output. Here, the graph is being maximized the more topics are being selected and increases once it reaches 100 topics.

In addition, Mimno et al. [MIM 11] propose a fifth measure, also dealing with coherence, that generally models with larger numbers of topics are preferred, as these models offer a higher resolution and the possibility of supporting finer-grained distinctions. Nevertheless, as the number of topics increases, the smallest topics can be of poor quality [MIM 11].

The last graph in Figure 10.6 clearly indicates that the topic quality after 100 topics decreases; hence, 100 topics should be chosen.

Regardless of the five aforementioned measures, in order to choose the "right" number of topics, a certain degree of human judgment is necessary. Hereby, it is important to notice that predictive accuracy and human judgment on the quality of topics are often not correlated [CHA 09], hence, for example, the measure of perplexity cannot be observed. After considering all the aforementioned measures, it has been decided to use two models, which are explained in the following.

In our example workflow, as indicated by Figure 10.1, our first model works with $K = 100$ topics corresponding with three keywords per topic. Alpha was defined with 0.5. Beta was defined with 0.001, which was also the predefined setting of KNIME. The second model, with $K = 50$ topics, corresponds to three associated keywords per topic. Alpha was defined with 1.0, and beta was defined with 0.001 once again. Therefore, 100, or respectively 50, major topics were identified by coding the keywords of the selected articles. In addition, it is noteworthy that previous studies in management literature, such as Furrer et al. [FUR 08], have been limited to only 26 keywords (see Table 1 in [FUR 08]). Table 10.1 once more illustrates the different parameters chosen.

Parameters	Model 1	Model 2
Topics (K)	100	50
Terms (T)	3	3
Alpha (α)	0.5	1.0
Beta (β)	0.001	0.001
Sensitivity	Coarse Granularity	Coarse Granularity

Table 10.1. *Model description with parameter selection*

The topic relevance is firstly assessed through the topic weightings given as an output measure by the LDA algorithm and then by a quantitative assessment based on the proximity of the originally published article keywords to the topic context in terms of a decadal and thematic (topic) assessment by an additional content analysis. The LDA algorithm makes use of the "*MALLET: A Machine Learning for Language Toolkit*" topic modeling library. MALLET is a Java-based package designed to deliver statistical natural language processing, document classification, clustering, topic modeling, information extraction and other machine learning applications to text [MCC 02].

Probabilistic topic modeling can be assigned to the branch of unsupervised machine learning algorithms aiming to annotate large achievements of documents with thematic information [BLE 12], i.e. to discover abstract topics in collections of unlabeled documents [KON 18]. Since the model is unsupervised, it does not require any a priori labeling of the documents. The topics then emerge from the analysis of the original texts in the form of a dimension reduction, which is similar to the numeric counterpart method of a principal component analysis (PCA), which is a simple, non-parametric method for extracting relevant information from cluttered datasets [PAU 13]. Nevertheless, LDA differentiates itself as it aims to reduce the information of textual (non-metric) data with the use of Bayesian statistics. LDA assumes that all collected documents share the same set of topics, but each document is exhibiting these topics in different proportions [BLE 12]. Speaking from a model-theoretic-based approach, a topic is the distribution over a fixed number of words, and it is assumed that these topics exist even before the generation of the data. LDA is a three-level hierarchical Bayesian model. Here, each word of a collection is being modeled as a finite mixture over a pre-existing set of topics. Each topic is modeled as an infinite mixture over a set of topic probabilities [BLE 03]. The described three-level process can be further explained as follows:

During step 1, each document exhibits all topics but in different proportions. Step 2 randomly selects a topic, which is chosen from the per-document distribution over topics. Afterwards, during step 3, every word in each document is being drawn from the previously chosen topic. To solidly ground the model's methodology, the probabilistic model is going to be described based on the preceding literature, such as Blei [BLE 12].

The stochastic problem can be attributed to the calculation of the posterior distribution, or the conditional distribution of the hidden (or latent) variables for the documents provided. Therefore, the LDA can be described through the following parameters.

The topics are described through $\beta 1{:}K$, while each βk describes the distribution over words. The topic proportions for the d-th document are θ_d, whereas $\theta d, k$ describes the topic proportion for topic k in document d. The topic assignments documents are z_d, whereas zd, n describes the topic assignment for the n-th word in document d. Finally, the observed words for each document d are wd, whereas wd, describes the n-th word in document d, which is a component from the fixed vocabulary that originates from the automatic read-out bag-of-words representation, which is a way of text data representation when text is being modeled through machine learning algorithms [BLE 12]. Therefore, the LDA process can be described, and the joined distribution of the hidden and the observed variables can be stated as follows [BLE 12]:

Equation 1: Joined distribution of hidden and observed variables

$$p(\beta_{1:K}, \theta_{1:D}, z_{1:D}, w_{1:D}) = \prod_{i=1}^{K} p(\beta_i) \prod_{d=1}^{D} p(\theta_d) \left(\prod_{n=1}^{K} p\left(z_{d,n} \mid \theta_d\right) p\left(w_{d,n} \mid \beta_{1:K}, z_{d,n}\right) \right)$$

Through this formula, numerous dependencies are being uncovered. According to this, the topic assignment zd, depends on the per-document proportions θd. Moreover, the observed word wd, n is dependent on both, the topic assignment zd, n as well as on all topics $\beta 1{:}K$ [BLE 12].

These dependencies define the LDA and are visualized in Figure 10.8. This figure shows how each node displays a random variable to its role in the generative process. Words within a document are characterized by shaded nodes. The rectangles denote replication. The *N* plate denotes the collection of words within the documents, while the *D* plate denotes the collection of documents within the collection [BLE 03]. Hidden variables are displayed by unshaded nodes, whereas observable variables are depicted by shaded notes. Rectangular boxes indicate multiple iterations.

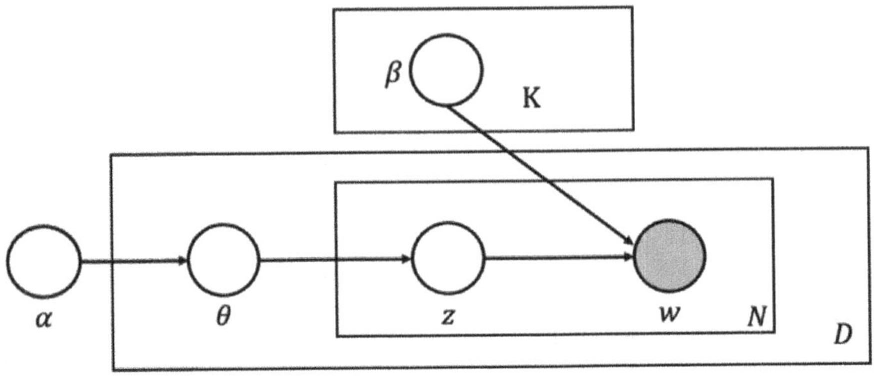

Figure 10.7. *The graphical model for LDA according to [BLE 12]*

The posterior probabilities can be calculated as follows:

Equation 2: Posterior probabilities

$$p(\beta_{1:K}, \theta_{1:D}, z_{1:D} \mid w_{1:D}) = \frac{p(\beta_{1:K}, \theta_{1:D}, z_{1:D}, w_{1:D})}{p(w_{1:D})}$$

In this formula, the numerator describes the joint variable of all random variables, precisely the topic distributions, document-specific topic proportions, topic assignment and word collection. The denominator describes the marginal probability of the words. Theoretically, they are computed through the aggregation of the joint distribution over every possible instance of the underlying hidden topic structure. Practically speaking, this distribution is difficult to calculate [DIC 83]. Therefore, a variational Bayesian framework as a means of an approximation method for the posterior probabilities is proposed. This approximation method assumes a Dirichlet distribution for each topic βk [ATT 01, KAM 21]. In addition, Blei et al. [BLE 03] suggest using another Dirichlet approximation. This approximation estimates the per-document topic proportions θd based on a fixed parameter called α (see Figure 10.8).

Since for this research two exemplary models have been created, the constant parameter differs, as previously described in this chapter. The constant parameters for model 1 (model 2) are $\alpha = 0.5$ ($\alpha = 1.0$) and $\beta = \beta k = 0.001$ ($= \beta k = 0.001$), and $K = 100$ (K = 50) has been chosen. Each topic is then associated with $T = 3$ major terms, whereas these terms describe the keywords indicating the highest probability that certain terms are associated with that one specific topic. As already mentioned, the alpha parameter defines the prior Dirichlet on the per-document topic distributions to the prior weight of the topics in a document. The KNIME library recommends using an alpha parameter of usually less than 1 for all topics to prefer sparse topic distribution, indicating few topics per document. The beta parameter defines the prior on per-topic multinomial distribution over words. Here, the KNIME library recommends using a number much smaller than 1; hence, the default value is set to 0.001. This serves the purpose to strongly prefer sparse word distributions. To further validate the robustness, a grid search optimization over the hyper-parameters α, β, K and T could be performed. Nevertheless, the grid search optimization presents itself as computationally intense; hence, a smaller variation analysis is proposed. This analysis aims to illustrate a first impression of how a successful grid search optimization could look like [KAM 21].

10.2.3. *Post LDA*

After the first nine nodes in the KNIME workflow, including the topic extraction through the LDA, have been completed, the next four nodes play a vital part for the processing and evaluation of the input data.

Topic id	Concatenate(Term)	Sum(Weight)	Detection
topic_00	customer, satisfaction, relationship	2584	Customer Satisfaction
topic_01	health, care, healthcare	892	Healthcare
topic_02	sharing, contracts, coordination	888	Contracting
topic_03	marketing, science, management	774	Marketing Science
topic_04	intertemporal, replication, formation	243	Intertemporal Choice
topic_05	capacity, management, project	1273	Project Management
topic_06	modeling, structural, model	998	Structural Modeling
topic_07	capital, human, venture	999	Human Capital
topic_08	simulation, estimation, variance	826	Simulation & Estimation
topic_09	goal, motivation, outsourcing	526	Outsourcing
topic_10	queue, systems, queueing	719	Service Optimization

Table 10.2. *Exemplary first 10 topics of the 100-topic model*

The first node (GroupBy) groups the rows of a table by the unique values in the selected group columns. For each unique set of values, a row is created within the selected group column. The columns which remain are aggregated and based on the specified aggregation settings. Finally, the output table contains one row for each unique value combination of the selected group columns. The next node is a "JavaSnippet", which allows us to execute arbitrary java code to create new columns or replace already existing ones. In this case, it renames the topics, from "topic_0" to "topic_00", for reasons of simplicity while dealing with the output. The next node, the "Sorter", sorts the rows according to the predefined criteria, meaning it sorts the topics starting from topic_00 to topic_99. Once these three nodes have been processed, the last node of this process is an "Excel writer". This node transfers the input data table into an Excel file, which directly downloads once it is being created by KNIME. Table 10.2 shows the output of the Excel writer, which is exemplary for the first 10 topics of the first model with 100 topics of this research.

At last, topic terms and weights have to be interpreted, and each topic is named by a qualitative assessment of the authors plus an additional two or more academics before they can be pivoted over the years. Once this list has been created, an extra column, named "Detection" has to be added next to the column illustrating the weight. The detection column specifies the topic for each row, for example, topic_00 consists of the term's customer, satisfaction and relationship; therefore, the assigned topic in this case can be interpreted as "Customer Satisfaction". Once this process has been completed for both models, 100 and 50 topics, the Excel list is once again put into an Excel reader (previously described in section 10.2.1) within the KNIME workflow. Once that list has been put into the KNIME workflow, several analyses can be performed, as to be described in section 10.2.1.

Furthermore, various robustness checks can be conducted to evaluate the quality of the resulting topic-term distributions, as already indicated in Figure 10.6. The association of terms to keywords and topics is analyzed qualitatively for an independently chosen subset of articles suggesting the economical soundness of the LDA clusters. In the following section, this research further presents an MCA analysis to quantitatively validate the concordance of article keywords and topic-term associations. Therefore, a two-component-based MCA can be used to project the underlying variation of data into two main factors. As our experience shows, an enlargement in the number of principal components does not generally lead to an increase in explained variation (regarding the maximum Eigenvalue criteria). Therefore, it can be assumed that the validation analysis with two main components is sufficient to test the suitability of the LDA projections [KAM 21].

Nevertheless, before turning to the results of the MCA and the content analysis, several limitations with regard to the proposed method can be depicted.

10.2.4. *Limitation of the LDA methodology and comments*

LDA makes three assumptions, which are going to be commented on in the following:

The first assumption of LDA is the "bag of words" representation, i.e. the order of the words in the document does not matter. This is problematic from a semantic viewpoint, especially when it comes to language generation. Nevertheless, this assumption does not limit analysis mainly consisting of stand-alone keywords.

The second assumption is that the order of the documents does not matter. As stated in the subsequent below, the focus in certain topics can shift over time. Thus, it may be reasonable to estimate multiple LDAs over time. By using this approach, a topic would be a sequence of distributions over words. To capture these dynamics, the author performed a content analysis that highlights the relevance of each topic over decades.

The last assumption about LDA is that the number of topics is to be assumed known and fixed. The Bayesian non-parametric topic model provides a solution for this [TEH 06]. Hereby, the number of topics is determined through the collection during posterior inference analysis and new documents can exhibit previously unseen topics.

10.3. Analyses

To begin with, it is worth knowing how many articles have been published in each year. If there are a large number of years to investigate, it may be beneficial to define different timespans. Figure 10.8 exemplifies the number of chosen articles over a four-year timespan for our example workflow. As it can be seen, the number of research articles in Management Science (comparably to other disciplines) has tended to rise in the recent decade. Note that the last time span declines since the collected articles were selected at the beginning of the year 2022. Thus, the overall collection pre-dominantly consists of articles from 2021 compared to a full four-year timespan of articles in previous periods.

For understanding, readability and simplicity purposes, only one model of the previously mentioned two is going to be used to explain the ongoing analyses as an additional part of our research methodology. An MCA is used to examine and analyze the relationship between the topic terms and keywords to gain a deeper insight into the relationships between the topics. In addition, the MCA can be used to evaluate the relationships between the topics over time [FUR 08].

MCA describes an exploratory data analysis technique used for graphical representation of multivariate categorical data [BEN 84, HOF 86, LEB 84]. This technique aims to explain the interdependence between a number of categorical variables – in this case, the topics (themes) and the keywords. It is similar to the PCA introduced by Hoffmann and Leeuw [LEE 92]. Through this technique, the researcher is being given the opportunity to examine and analyze multipath tables, and determine the structure in the relationship between the nominal variables

[FUR 08]. Through this analysis, similar counting patterns between the different rows and columns of a frequency table can be highlighted.

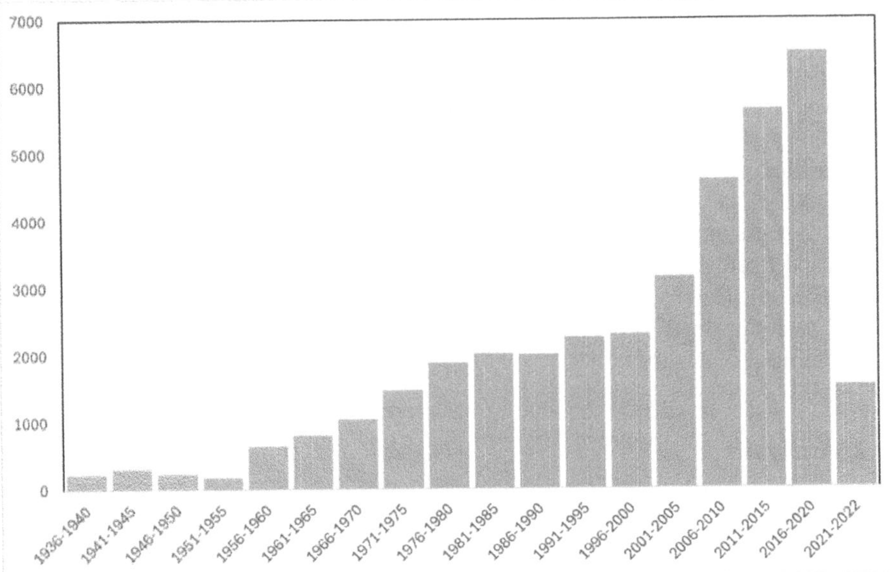

Figure 10.8. *Number of articles per timespan*

Figure 10.9 shows that the collected keywords from all articles (blue dots) of the example data overlap perfectly with the associated terms (green dots) and subordinate topics (orange dots). This outcome supports the use of the LDA topic extractor, which has been previously described.

For this purpose, a two-dimensional factor has been determined as the most suitable option for the graphical representation. The corresponding eigenvalues were both around one, with an explained inertia of 0.02% for each component.

As Kaciak and Louviere [KAC 90] noted, the proportion of total variance explained by the dimensions is often very small. This is closely related to the (necessarily) binary nature of the transformed nominal data [LEB 84]. Additional sensitivity analyses can be carried out for higher orders of components ($n = 2 \dots 10$), which did not lead to any improvement in the maximum explained inertia per component in this example.

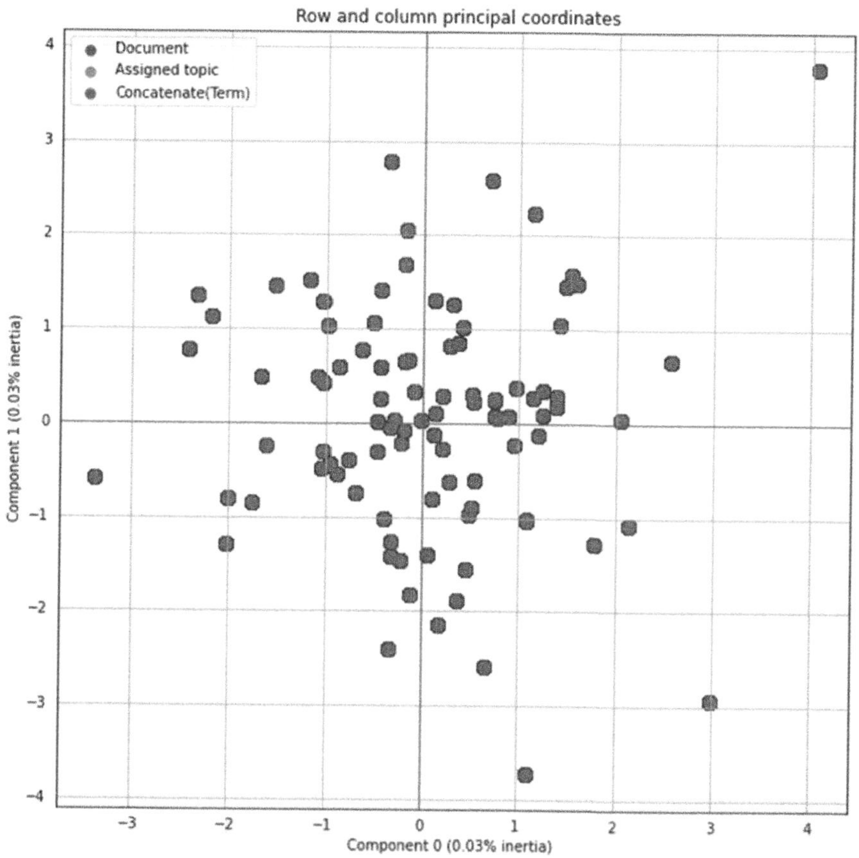

Figure 10.9. *Exemplary MCA analysis illustrating the proximity of observable research article keywords (blue dots), topic assignments (orange dots) and (concatenated) term assignment (green dots), with the dot size indicating the number of articles associated with the categories*

In general, the MCA results illustrated above prove a high match between the observed literature articles and the topic-term distributions. In subsequent analyses, researchers can evaluate advanced visualization methods, which are included in the Python library LDAvis developed by Sievert et al. [SIE 14]. The authors propose a new relevance score in order to assess the degree to which a term belongs to a topic by using a fraction of the log-likelihood of the term $w_{d.n}$ and the approximated Bayesian distribution over the empirical term frequency (as introduced in Equation 1 of section 10.2.2).

Previous studies such as Bischof and Airoldi [AIR 12] simply ranked the terms by the probability under a topic. This inadequacy is being overcome by the proposed measure in this methodology. In the previous section, a different visualization, comparable to Sievert et al.'s [SIE 14] method, by using an MCA between topic term distributions, has been used. In addition, to highlight the changes of topics over time, a content analysis can be performed, distinguishing between topic and year associations.

10.4. Results of the content analysis

The exemplary content analysis depicted in Figure 10.10 was executed to illustrate the relationships between the assigned topics and the grouped years in our example. Through a pivot table, which was created after the LDA extraction from the assigned topics over years, it was determined how frequently a published article appeared within each topic over each year group. Once again, to simplify the graphical representation, a two-factor model has been chosen as the most suitable option. The blue bubbles describe the topics, while the orange bubbles describe the time period. The size of the orange bubbles indicates the number of articles associated with the respective timespan, while the size of the blue bubbles indicates the number of articles associated with each topic. The closer the year and topic bubbles are together, the closer their association in the sense of thematic proximity. This is also the case from a geometrical viewpoint, in accordance with the Euclidean distance, which is the length of a line segment between the two points. The size of the topic bubbles describes the number of articles published on that topic; hence, larger topic bubbles have more articles associated with them. The larger the size of the year bubble, more articles have been assigned to that specific time period, and hence, more articles have been published during that time period.

A pair of coordinates in a two-dimensional space is provided by the analysis for all articles in the exemplary input dataset. Nevertheless, the map would be impossible to interpretate if a dot was presented for each article, and therefore, only the position of the topics is being indicated. Moreover, the MCA described earlier implicates that the article keywords and the assigned topics are closely correlated.

Furthermore, numerous time series analyses can be performed to marginally assess the evolution of a single topic over time; a small excerpt of examples is presented in Figure 10.11. A positive, respectively negative, slope of the trend line (red line) indicates an increase or decrease in the average amount of articles that can be associated with each topic throughout the timespan of analysis. The time series plots underline the relevance of the topics over the years, proving a constant increase or a high number of publications constantly throughout the years or, respectively, the opposite.

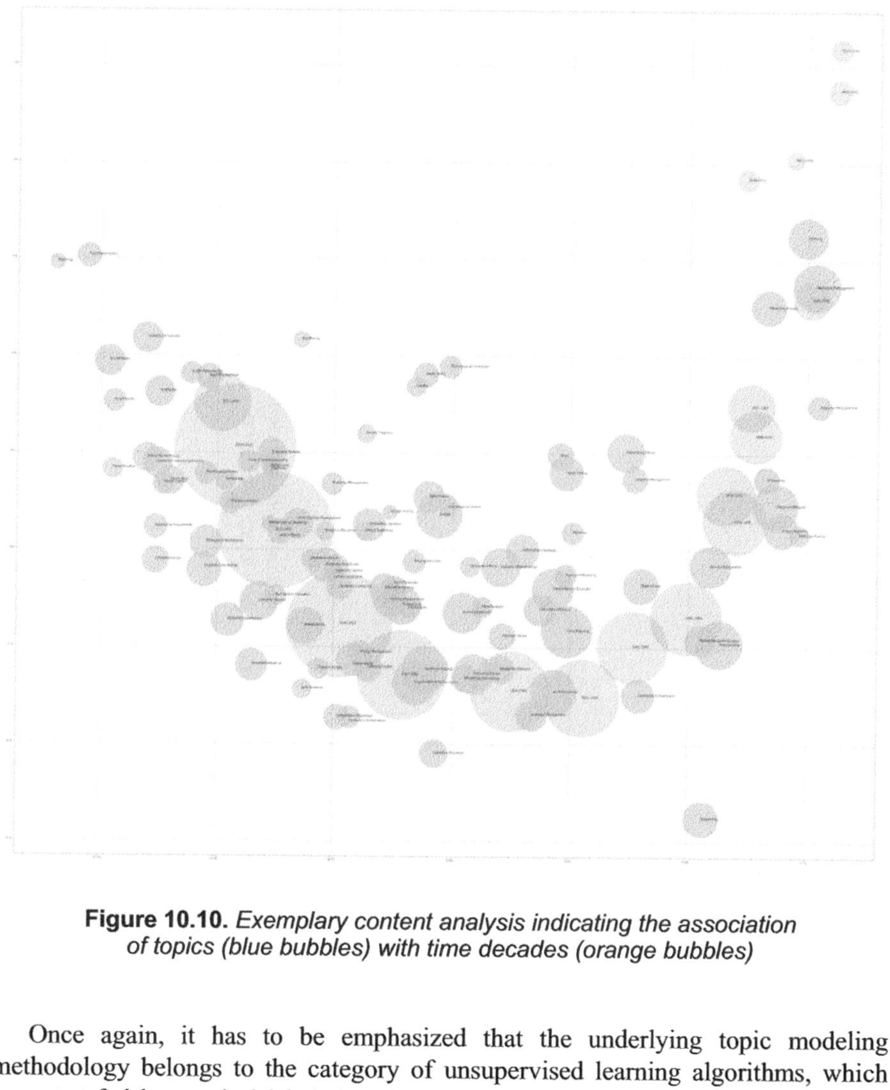

Figure 10.10. *Exemplary content analysis indicating the association of topics (blue bubbles) with time decades (orange bubbles)*

Once again, it has to be emphasized that the underlying topic modeling methodology belongs to the category of unsupervised learning algorithms, which were not fed by a priori labeled topic or term names, as can be seen in previous literature studies such as Furrer et al. [FUR 08].

The algorithm simply groups keywords that are close in topic, which, as the previous robustness checks have shown, seems reasonable from a management perspective. This overcomes subjective topic labeling by researchers embedded in a particular research environment, which could lead to biased topic evaluations.

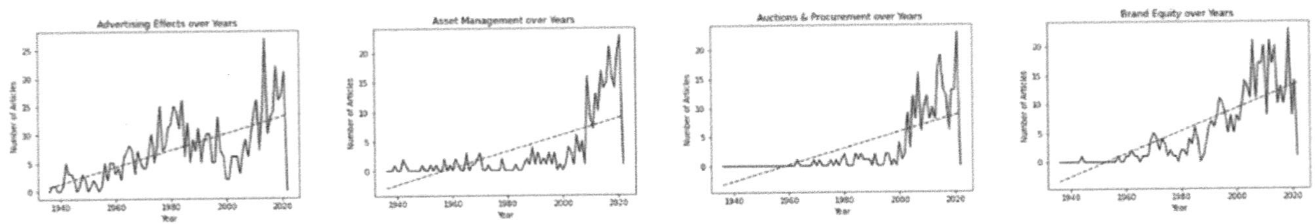

Figure 10.11. *Exemplary time series plots (blue) with trendlines (red) showing the evolution of articles per topic over time*

Figure 10.12. *Content analysis indicating the association of topics (blue bubbles) with journals (orange bubbles)*

In addition to the assignment of topics to a specific timespan, the analysis can also output the following (Figure 10.12). Here, the topics are being assigned to the journals analyzed, indicating which focus each journal has. This offers a particularly interesting insight, especially for researchers, regarding new publications and where they could best submit their newest publications based on their topic. For example, researchers focusing on "Logistics Management", "Programming" or "Scheduling" have a high probability of being published with one of those topics in the *Journal of Operations Research*, while researchers focusing on "Strategic Management", "Innovation Alliances" or "Internationalization" will likely be rejected by the *Journal of Operations Research*, but have a better chance of being published in the *Strategic Management Journal*. Similar to Figure 10.11, which assigned the topics to a specific timespan, this illustration also presents topics, which cannot be precisely assigned to one journal since their overlap with the topics represented in the journals is too marginal. Namely, for example, these topics are, "Game Theory", "Gender", "Technological Innovation" or "Purchase Management" to only name a few in the example dataset. These topics are additionally represented in relatively small bubbles simply because not enough articles could be associated with these topics, and therefore, the assumption arises that they could not be assigned to a specific journal.

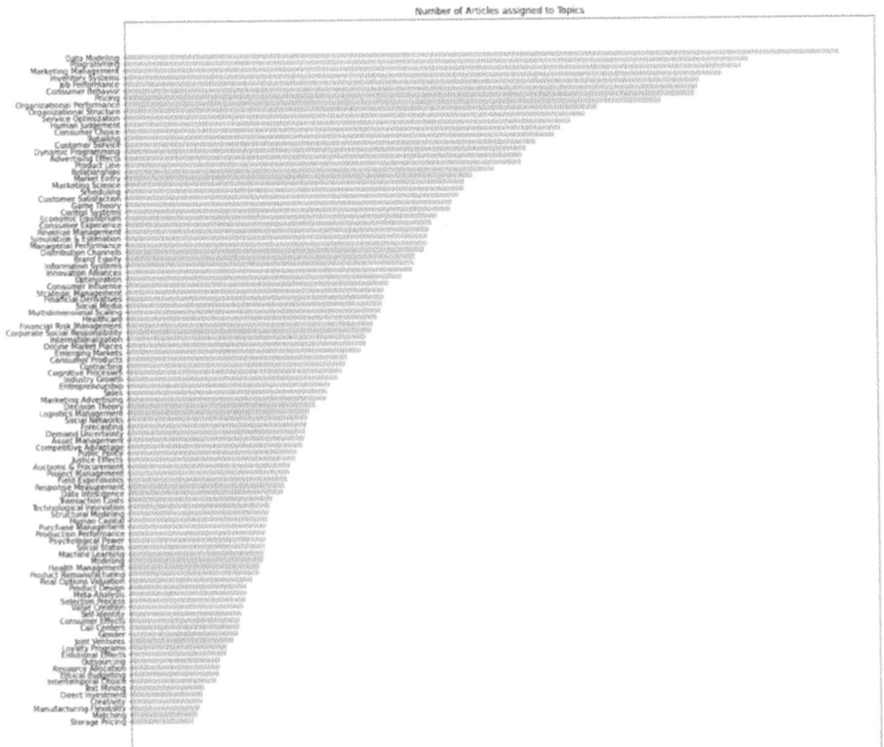

Figure 10.13. *Number of articles assigned to topics*

Once again, the size of the orange bubbles indicates the number of articles associated with, in this case, each journal, while the size of the blue bubbles indicates the number of articles associated with each topic. The closer the bubbles are together, the higher their correlation. Therefore, this figure allows researchers, aiming to publish an article, to pick the right journal precisely fitted to their research field.

Figure 10.13 illustrates the number of articles associated with each extracted topic of our exemplary workflow (with $K = 100$ topics). In this example, there was a large group of research articles associated with the topics of "Data Modeling", "Programming" and "Inventory Systems", underpinning once again the impact of digitalization and automation issues in Management Science.

10.5. Contributing authors

Bergh et al. [BER 06] concluded that the characteristics of each author have the greatest explanatory power regarding the impact of the article. They implied that those researchers, who published the most within a certain period, subsequently had the strongest influence on topics researched in the subsequent period. For the purpose of this research, the authors with the most publications throughout the time span of publication of all articles are proposed to be analyzed to fully understand the history of the management field and allow for possible future implications and topics. Moreover, Berry and Parasuraman [BER 93] proposed a model regarding the forces that influence the development of a certain academic field. Hereby, key individuals (authors) and key publications act as accelerating factors for each academic field. Therefore, the most published authors, as well as the most published papers with regard to their impact, can be analyzed with the proposed methodology [FUR 08].

Through this analysis, the contributions of researchers over the time period analyzed are highlighted, differentiating between long-established scholars as well as new generations of scholars. New scholars will face the task of determining future research and the direction the field of management is heading in. The number of published papers by an author runs proportionally to their career length; hence, the authors on top of the list have the longest career, and therefore, their influence on the development of the management field can be regarded as the highest.

10.6. Most influential papers

Specific publications can be assigned with a vital and central role regarding the growth of the scientific field they were published in, since their influence accelerated the growth of the discipline. In order to understand the future possibilities and capabilities of the management discipline from a holistic perspective, it is necessary to detect and identify the major papers published in all journals analyzed for the respective time span. Therefore, the approach of summed citation counts can be used to detect and assess their impact or influence on research papers [BER 06, RAM 04, TAH 99].

The most cited articles of the respective journals can be determined based on the citation count they have been assigned in the Web of Science. Since papers, which have been published at an earlier point in time, have a greater chance of receiving a higher citation count than papers published at a later stage in time, the articles have to be sorted by the number of citations divided by the number of years they have been published. The procedure proposed by Furrer et al. [FUR 08] can therefore be used.

10.7. Box plotting

In addition to the determination of the most published authors and the most influential papers, regarding their citations per year, several boxplots for each topic can be determined. This serves to determine whether a topic a researcher is writing about leads to a potentially high citation count of the respective article.

Boxplots are diagrams used to graphically represent the distribution of an at least ordinally scaled characteristic. An ordinal scale sorts variables with expressions between which there is a ranking order. It thereby combines various robust measures of dispersion and position in one representation. A boxplot is intended to quickly give an impression of the range in which the data lies and how they are distributed over this range [KRO 14]. Figure 10.14 shows an exemplary structure of a boxplot.

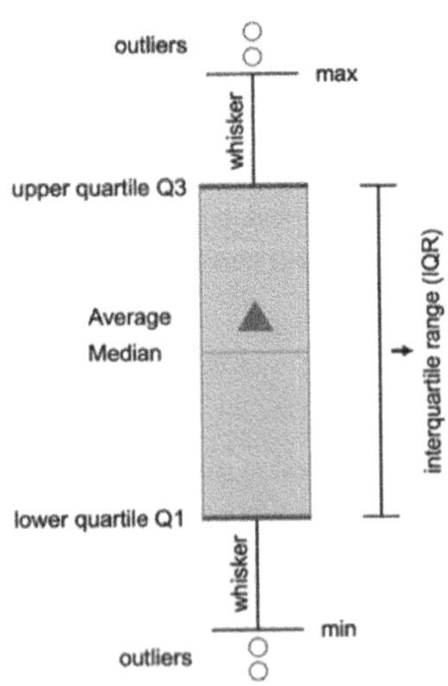

Figure 10.14. *Exemplary boxplot structure*

The lowest observed score, excluding the outliers, is shown at the end of the bottom whisker. In total, 25% of all scores are illustrated below the first, while 75% of all scores are illustrated below the upper quartile; hence, 25% of all scores are above that value. The median marks the mid-point of the analyzed data and divides the box into two parts. Similar to the lowest observed score, the highest score, excluding outliers, is shown at the end of the top whisker. The whiskers illustrate the scores that are outside the middle 50%. The interquartile range simply illustrates the middle 50% of all scores. The outliers can be seen as an observation, which is numerically distant from the rest of the observed data. These outliers are located outside the whiskers [MCL 19]. The float determines the reach of the whiskers beyond the first and third quartiles. In other words, where IQR is the interquartile range (Q3 – Q1), the upper whisker extends to the last datum less than Q3 plus the whisker times the IQR (or briefly 1.5*IQR). Similarly, the lower whisker will extend the first datum greater than Q1 minus the whisker times the IQR (or briefly 1.5*IQR). The outliers are plotted as individual points. In the Python code, this value is set unreasonably high to force the whiskers to show the minimum and maximum values.

The higher the boxes, the higher the probability that an article about the respective topic is being cited; hence, the lower the box, the lower the probability of reaching a high citation count with that respective topic. The higher the value on the vertical axis, the higher the citation count of the respective topic. If the median and the average are on top of each other or at least close together, the topic is more symmetrically distributed.

An additional boxplot analysis of only the most cited articles per journal can be performed to be compared to the results of the total dataset. The two boxplot analyses give a deeper insight into which topics offer the highest opportunity to be cited for researchers.

10.8. Conclusion

In this study, the author has presented a comprehensive methodology for topic modeling and analysis of management research publications. By applying LDA to extract topics, this research successfully revealed latent thematic structures within the corpus. The post-LDA processing facilitated the visualization and interpretation of topic-term distributions, contributing to a better understanding of key research themes.

The application of MCA provided quantitative validation, confirming the alignment between article keywords and topic assignments. Through content analysis and time series plots, this research delved into the evolution of topics over time, highlighting persistent areas of interest and emerging trends in management research.

By examining the most influential authors and papers based on citation counts, the author acknowledged the significant contributions of key individuals and landmark publications in shaping the management field. Moreover, the boxplot analysis allowed us to identify topics with higher potential for citations, aiding researchers in making informed decisions when selecting research topics.

Overall, our methodology offers valuable insights into the management research landscape, aiding scholars, practitioners and journal editors in understanding and navigating the diverse and ever-evolving field of management. As the body of management literature grows, the utilization of topic modeling and analysis becomes increasingly important in facilitating knowledge discovery and fostering new avenues of research.

10.9. References

[ARU 10] ARUN R., SURESH V., VENI MADHAVAN C.E. et al., "On finding the natural number of topics with latent Dirichlet allocation: Some observations", in ZAKI M.J., YU J.X., RAVINDRAN B. et al. (eds), *Advances in Knowledge Discovery and Data Mining, Part I. 14th Pacific-Asia Conference, PAKDD 2010, Hyderabad, India, 21–24 June 2010, Proceedings*. Springer, New York, pp. 391–402, 2010.

[ATT 01] ATTIAS H., PLATT J.C., ACERO A. et al., "Speech denoising and dereverberation using probabilistic models", *Advances in Neural Information Processing Systems*, available at: https://papers.nips.cc/paper_files/paper/2000, 2001.

[BEN 84] BENZÉCRI J.-P., BELLIER L., *L'analyse des données. Leçons sur l'analyse factorielle et la reconnaissance des formes et travaux du laboratoire de statistique de l'Université de Paris VI*, 4th edition, Dunod, Paris, 1984.

[BER 93] BERRY L.L., PARASURAMAN A., "Building a new academic field—The case of services marketing", *Journal of Retailing*, vol. 69, no. 1, pp. 13–60, 1993.

[BER 06] BERGH D.D., PERRY J., HANKE R., "Some predictors of SMJ article impact". *Strategic Management Journal*, vol. 27, no. 1, pp. 81–100, 2006.

[BIS 12] BISCHOF J.M., AIROLDI E.M., "Summarizing topical content with word frequency and exclusivity", in *Proceedings of the 29th International Conference on Machine Learning*, Edinburgh, 2012.

[BLE 03] BLEI D.M., NG A.Y., JORDAN M.I. et al., "Latent dirichlet allocation", *Journal of Machine Learning Research*, vol. 3, pp. 993–1022, 2003.

[BLE 12] BLEI D.M., "Probabilistic topic models", *Communications of the ACM*, vol. 55, no. 4, pp. 77–84, 2012.

[CAO 09] CAO J., XIA T., LI J. et al., "A density-based method for adaptive LDA model selection", *Neurocomputing*, vol. 72, nos 7–9, pp. 1775–1781, available at: https://www.sciencedirect.com/science/article/pii/S092523120800372X, 2009.

[CHA 09] CHANG J., BOYD-GRABER J.L., GERRISH S. et al., "Reading tea leaves: How humans interpret topic models", in *Advances in Neural Information Processing Systems 22: 23rd Annual Conference on Neural Information Processing Systems 2009. Proceedings of a Meeting held 7–10 December 2009, Vancouver, British Columbia, Canada*, available at: https://www.researchgate.net/publication/221618226_Reading_Tea_Leaves_How_Humans_Interpret_Topic_Models, 2009.

[DIC 83] DICKEY J.M., "Multiple hypergeometric functions: Probabilistic interpretations and statistical uses", *Journal of the American Statistical Association*, vol. 78, no. 383, p. 628, available at: http://www.jstor.org/stable/2288131, 1983.

[ELL 06] Ellen P.S., "Building corporate associations: Consumer attributions for corporate socially responsible programs", *Journal of the Academy of Marketing Science*, vol. 34, no. 2, pp. 147–157, available at: https://link.springer.com/article/10.1177/0092070305284976, 2006.

[FUR 08] FURRER O., THOMAS H., GOUSSEVSKAIA A., "The structure and evolution of the strategic management field: A content analysis of 26 years of strategic management research", *International Journal of Management Reviews*, vol. 10, no. 1, pp. 1–23, 2008.

[HOF 86] HOFFMAN D.L., FRANKE G.R., "Correspondence analysis: Graphical representation of categorical data in marketing research, *Journal of Marketing Research*, vol. 23, no. 3, p. 213, available at: http://www.jstor.org/stable/3151480, 1986.

[HOF 92] HOFFMAN D.L., LEEUW J., "Interpreting multiple correspondence analysis as a multidimensional scaling method", *Marketing Letters*, vol. 3, no. 3, pp. 259–272, available at: https://www.academia.edu/2611242/Interpreting_multiple_correspondence_analysis_as_a_multidimensional_scaling_method, 1992.

[HUB n.d. a] HUB.KNIME.COM, Excel Reader, available at: https://hub.knime.com/knime/extensions/org.knime.features.ext.poi/latest/org.knime.ext.poi3.node.io.filehandling.excel.reader.ExcelTableReaderNodeFactory [Accessed 01 April 2022].

[HUB n.d. b] HUB.KNIME.COM, Strings To Document, available at: https://hub.knime.com/knime/extensions/org.knime.features.ext.textprocessing/latest/org.knime.ext.textprocessing.nodes.transformation.stringstodocument.StringsToDocumentNodeFactory2 [Accessed 01 April 2022].

[JAC 16] JACOBI C., VAN ATTEVELDT W., WELBERS K., "Quantitative analysis of large amounts of journalistic texts using topic modelling", *Digital Journalism*, vol. 4, no. 1, pp. 89–106, available at: https://www.researchgate.net/publication/283671339_Quantitative_analysis_of_large_amounts_of_journalistic_texts_using_topic_modelling, 2016.

[KAC 90] KACIAK E., LOUVIERE J., "Multiple correspondence analysis of multiple-choice experiment data", *Journal of Marketing Research*, vol. 27, no. 4, pp. 455–465, 1990.

[KAM 23] KAMRAN Q., TOPP S., BECKER M., "The structure and evolution of the marketing field: A content analysis of five decades of research within the Academy of Marketing Science Journals", *Research Journal for Applied Management (RJAM), International School of Management (ISM), Dortmund*, vol. 4, no. 1, pp. 89–132, 2023.

[KON 18] KONRAD M., Probabilistic topic modeling with LDA. Practical topic modeling: preparation, evaluation, visualization, Python User Group Workshop, available at: http://dsspace.wzb.eu/pyug/topicmodeling2/slides.html [Accessed 01 Apri 2022], 2018.

[KRO 14] KRONTHALER F., *Statistik angewandt. Datenanalyse ist (k)eine Kunst*, Springer Spektrum, Berlin, Heidelberg, 2014.

[LEB 84] LEBART L., MORINEAU A., WARWICK K.M. *Multivariate Descriptive Statistical Analysis: Correspondence Analysis and Related Techniques for Large Matrices*, Wiley, New York, 1984.

[MAI 04] MAIGNAN I., FERRELL O.C., "Corporate social responsibility and marketing: An integrative framework", *Journal of the Academy of Marketing Science*, vol. 32, no. 1, pp. 3–19, available at: https://link.springer.com/article/10.1177/0092070303258971, 2004.

[MCC 22] MCCALLUM A.K., MALLET: A Machine Learning for LanguagE Toolkit, available at: http://mallet.cs.umass.edu [Accessed 1 April 2022], 2022.

[MCL 19] MCLEOD S., Box plots (also known as box and whister plots), available at: https://www.simplypsychology.org/boxplots.html [Accessed 1 April 2022], 2019.

[MIM 11] MIMNO D.M., WALLACH H.M., TALLEY E.M. et al., "Optimizing semantic coherence in topic models", in *Proceedings of the 2011 Conference on Empirical Methods in Natural Language Processing, EMNLP 2011*, 27–31 July, John McIntyre Conference Centre, Edinburgh, pp. 262–272, available at: https://www.researchgate.net/publication/221012637_Optimizing_Semantic_Coherence_in_Topic_Models, 2011.

[NEW 09] NEWMAN D., ASUNCION A.U., SMYTH P. et al., "Distributed algorithms for topic models", *Journal of Machine Learning Research*, vol. 10, pp. 1801–1828, available at: https://www.researchgate.net/publication/220320734_Distributed_Algorithms_for_Topic_Models, 2009.

[PAU 13] PAUL L.C., AL SUMAN A., SULTAN N., "Methodological analysis of principal component analysis (PCA) method", *International Journal of Computational Engineering & Management*, vol. 16, no. 2, pp. 32–38, available at: https://researchers.mq.edu.au/en/publications/methodological-analysis-of-principal-component-analysis-pca-metho, 2013.

[RAM 04] RAMOS-RODRÍGUEZ A.-R., RUÍZ-NAVARRO J., "Changes in the intellectual structure of strategic management research: A bibliometric study of the Strategic Management Journal, 1980–2000", *Strategic Management Journal*, vol. 25, no. 10, pp. 981–1004, 2004.

[RÖD 15] RÖDER M., BOTH A., HINNEBURG A., "Exploring the space of topic coherence measures", in *WSDM '15: Proceedings of the Eighth ACM International Conference on Web Search and Data Mining*, pp. 399–408, doi: 10.1145/2684822.2685324, 2015.

[SCI n.d.] SCIMAGOJR.COM, Scimago Journal & Country Rank, available at: https://www.scimagojr.com/ [Accessed 1 April 2022].

[SEN 06] SEN S., "The role of corporate social responsibility in strengthening multiple stakeholder relationships: A field experiment", *Journal of the Academy of Marketing Science*, vol. 34, no. 2, pp. 158–166, available at: https://link.springer.com/article/10.1177/0092070305284978, 2006.

[SIE 14] SIEVERT C., SHIRLEY K., "LDAvis: A method for visualizing and interpreting topics", in *Proceedings of the Workshop on Interactive Language Learning, Visualization, and Interfaces. Proceedings of the Workshop on Interactive Language Learning, Visualization, and Interfaces*, Association for Computational Linguistics, Baltimore, MD, pp. 63–70, 2014.

[TAH 99] TAHAI A., MEYER M.J., "A revealed preference study of management journals' direct influences", *Strategic Management Journal*, vol. 20, no. 3, pp. 279–296, available at: http://www.jstor.org/stable/3094106, 1999.

[THE 06] TEH Y.W., JORDAN M.I., BEAL M.J. et al., "Hierarchical Dirichlet processes", *Journal of the American Statistical Association*, vol. 101, no. 476, pp. 1566–1581, https://www.researchgate.net/publication/221997021_Hierarchical_Dirichlet_Processes, 2006.

[TUF 17] TUFTS C., *The Little Book of LDA* [work in progress], available at: https://miningthedetails.com/LDA_Inference_Book/ [Accessed on 1 April 2022], 2017.

[TUT 21] TUTORIALSPOINT.COM, OpenNLP – Tokenization, available at: https://www.tutorialspoint.com/opennlp/opennlp_tokenization.htm [Accessed 1 April 2022], 2021.

[YAO 09] YAO L., MIMNO D., MCCALLUM A., "Efficient methods for topic model inference on streaming document collections", in *KDD'09: Proceedings of the 15th ACMKDD International Conference on Knowledge Discovery & Data Mining*, 28 June–1 July, Paris. ACM, New York, pp. 937–946, 2009.

10.10. Appendix 1. Application of the machine learning methodology to investigate the domain of entrepreneurship and marketing

TITLE: DESIGN FOR A SERIAL ENTREPRENEURSHIP IN CYBER-PHYSICAL REALITIES – DECIPHERING VISIONARY MANAGEMENT MODELS FOR MARKETING

This research is developed as a case to analyze and develop a holistic literature review of 21 leading journals in management science from 1960 to 2022 by using unsupervised machine learning, namely KNIME and Python, to determine the evolutionary history of the fields and to decipher the foundations of integrative models of serial entrepreneurship enhancing entrepreneurially driven marketing strategies.

Keywords: literature review, entrepreneurially driven marketing, serial entrepreneurship, interdisciplinary Weltanschauung of marketing, KNIME

Description: This application of methodology provides an overview of entrepreneurship, innovation, leadership and cybernetics by analyzing 36,348 articles published in 21 leading journals from 1960 until January 2022 to develop a new conceptual framework supporting marketers to operate in more complex and dynamic markets.

10.10.1. *Introduction*

Especially in recent years, the world has been marked by unexpected crises, such as the Covid-19 pandemic and the Russian invasion of Ukraine, significantly changing the business environment. While these crises are still lingering and have dramatically increased the range of entrepreneurial and marketing pursuits, the enduring spirit of creativity and perseverance will address these challenges. Therefore, a design for serial entrepreneurship to decipher visionary management models delivering profound recommendations to marketers will be delivered. By applying design thinking as the methodology of this chapter, it aims first to point out the status quo in entrepreneurship, leadership, innovation and cybernetics. In a second step, the holistic second-order entrepreneurial Design Weltanschauung model (SOEDWAM) will be developed. Our framework is based on an interdisciplinary Weltanschauung of marketing. It combines the sciences of entrepreneurship, cybernetics and philosophy into a coherent whole, thus helping marketers find their way in complex marketing situations.

10.10.2. *Method*

The author used a Latent Dirichlet Allocation (LDA) algorithm for unsupervised topic modeling and Python as an additional tool for the following analyses. Several content analyses (CAs) must be examined to connect the existing managerial literature with cybernetic journals upon which the SOEDWAM for serial entrepreneurship in cyber-physical realities can be built. This is because a few latent topics are enough to represent a large corpus effectively. According to [ARU 10], such models have proved very effective. The related multiple correspondence analysis (MCA) was conducted to validate the mapping between the digitized keywords from 36,348 research articles and the topic labels and terms resulting from the performed LDA analysis. In this way, the scientific analysis of the "Lay of the Land" in innovation, entrepreneurship, leadership and cybernetics with the techniques and tools of artificial intelligence were combined to achieve more precise and objective labeling of the "topic-keyword relationships". LDA assumes that each topic is associated with multiple terms, and conversely, terms may belong to multiple topics but have different effects on the overall topic distribution. Based on current developments in ML capabilities, this methodology provides a solid foundation not only for this research, but also for future approaches where a large amount of textual data is to be analyzed for a more holistic grounded theory design. A holistic view presents the most discussed past, present and future topics within the main research areas. Using unsupervised machine learning to analyze the massive number of articles in combination with the mixed research approach, namely quantitative and qualitative, ensures the robustness and applicability of the model referring to the most essential literature in the field. Table 10.1 visualizes the method of the analysis.

Data Import	Pre-Processing	Topic Detection via LDA	Grouping	Visualization
• Selection of 35,840 / 36,348 / 33,206 articles (21 top journals in entrepreneurship, innovation, leadership, and cybernetics) • Creation of relevant Excel list including all relevant information (authors, title, journal, keywords, abstract, citations, citations per year)	• Coding of keywords to generate 100 major topics per hand • Tokenization of information about title, authors, and keywords • Punctuation erasure • Filter out numbers • Delete words with less than 3 characters • Creation of stop list to exclude specific common words	• Single-parallel distributed implementation of LDA in KNIME for detecting the topics most relevant to the research project • The node uses the MALLET topic modeling library • Generation of 100 topics based on sensitivity analysis of 5 terms each • alpha = 0.01 and Beta = 0.001=> fine granularity	• Grouping of topic terms and weighting with qualitative evaluation • Pivoting of topics over years	• MCA was applied to identify the relationship between topic terms and keywords • MCA on the relevant journals for this research project • Time series plots with trendlines show assigned articles per topic over years • CA-analyses to present the interrelation of assigned topics to grouped years and journals • Box Plots
Main Interpretations Resulting from the analyses				
• MCA-Analyses indicate accurate results regarding the applied methodology • Entrepreneurial and cybernetical topics are identified as the most discussed topics in present and possible future due to the trend lines • Substantial scientific gap between cybernetics and management will be closed by this thesis				

Table 10.3. *Method of the analysis*

The data analysis software "KNIME" was used for the analyses, a commercial API software with an AI-based drag-and-drop solution. A simple parallel distributed implementation of LDA by [NEW 09] using the sparse LDA sampling scheme and the data structure of [YAO 09] was used to identify the most relevant topics in the existing literature related to entrepreneurship, innovation, leadership and cybernetics over the last seven decades. Unsupervised machine learning methods assessed the 100 most essential topics in these disciplines, and each article was assigned to one precise topic.

Figure 10.15. *CA-Analysis, including cybernetics, indicates the association of topics (blue bubbles) with time decades (orange bubbles). The size of the bubbles indicates the number of associated articles for either the topic or year category, respectively. Bubbles closer together indicate thematic proximity induced by the underlying LDA classifier*

To show the interrelation of the assigned topics to grouped years, various CA Analyses were performed. The frequency of published articles within a topic was based on a pivoting table drawn from the assigned topics over the years after LDA extraction.

The author chose a two-factor model to simplify the graphical representation. Topics (blue bubbles) that are closely related to a period (orange bubbles) are also geometrically (in terms of the Euclidean distance) close together. The bigger the topic bubble, the more articles were published on the topic, and the bigger the journals' bubble, the more articles were published in the respective journals.

10.10.3. *CA – the most cited articles per journal per year*

In a second step, the author analyzed which of the 100 labeled topics occurred in the 840 most cited articles. The 40 most cited articles per journal per year, resulting in 840 articles, are categorized into 10 distinct topics, which are built by clustering the 100 most essential topics in management science, such as cybernetics, entrepreneurship, innovation, international management, leadership, marketing, operations research and strategy in consultation with an expert in the managerial field.

As the scope of this article is limited, not all 840 articles can be described in-depth. To give a short overview of the topic distribution regarding timespans and the categories that have been discussed in the essential literature since 1960, Figure 10.16 is shown. The bigger the diameter, the more articles were published in the respective category and decade. It indicates the importance of cybernetics and entrepreneurship nowadays and the constant importance of marketing throughout the 62-year period.

Design Weltanschauung model

By working through existing models and literature in Design Science, the Design Weltanschauung model developed by [KAM 21] has been detected, indicating the highest fit for this research to build upon. Operating in cyber-physical realities with global crises such as the Covid-19 pandemic indicates that businesses operate in markets shifting towards the complexity of massive proportions [KAM 20]. The second axis is the time to react to upcoming crises and challenges, which periodically becomes shorter [KAM 21]. Until an adequate strategy is formulated [MIN 20], the crises could become even more severe because the situation could have already changed significantly [BEE 93].

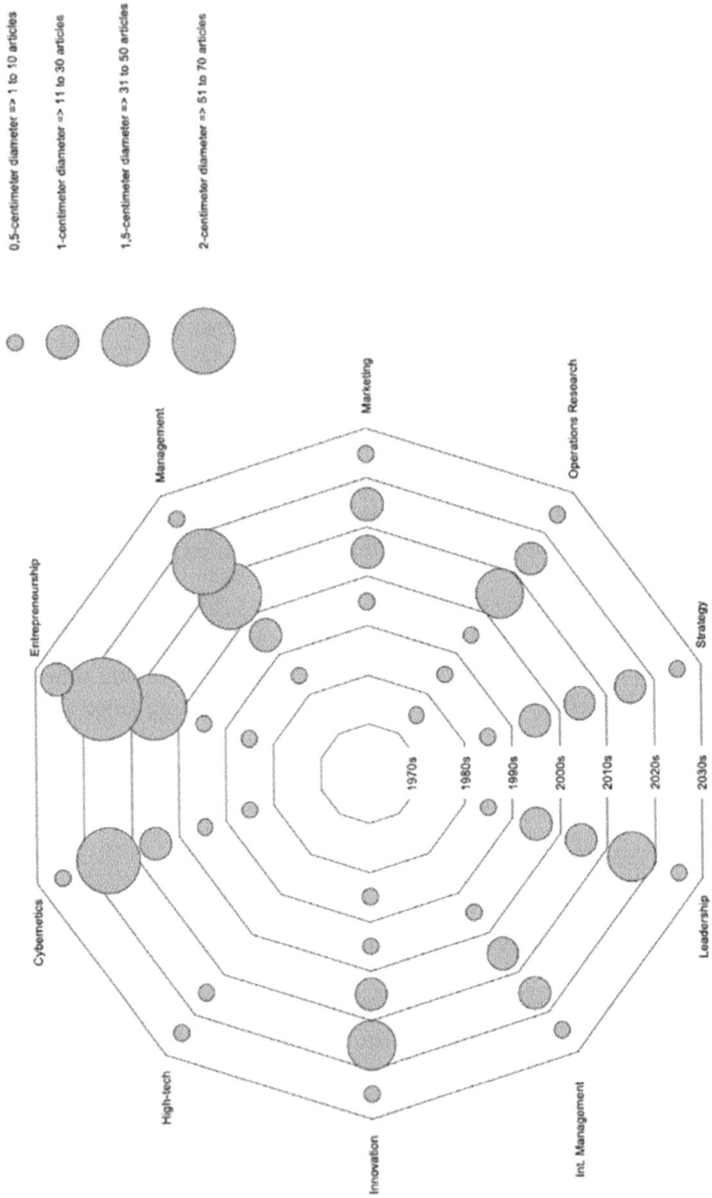

Figure 10.16. *840 most cited articles per year assigned to category and timespan. Deciphering visionary management models via the SOEDWAM*

Building upon the Design Weltanschauung Model model, marketers must ask themselves three key questions: (1) Who are the authors, and what is our legitimacy? (2) What does the author need to do today? (3) What does the author need to do in the future? [KAM 21]. The essential component of this model for SOEDWAM is the differentiation between the future and Jacques Derrida's analogy of L´avenir.

The future that will comparably surely come is predicted and foreseeable. It is considered the future of controlled, handled and programmed timetables. On the other hand, L´avenir is totally unexpectable and unpredictable, signaling the otherness of the future and the coming arrivant [TRE 19].

L´avenir is not an event with a beginning and ending point, and cannot be pinned down. L´avenir will always escape philosophy and mastery [FRE 22]. However, the situation does not mean that the author is powerless and incapable; the aporia is not the end of the road but the beginning. While the author cannot willfully actualize or make an event true, the author can make themselves available to it. The author can be true to the events [TRE 19].

L´avenir is that which distinguishes or unhinges the movement in question [MAR 07]. The analogy describes an individuum or event arriving without the competitor being able to anticipate the arrival [DIC 05]. The difference between le futur and L'avenir, therefore, is what the future does or what the author does with the future (L'avenir) over what the future is or holds (le futur) [MAR 07].

L´avenir is a certain impossibility, an aporetic structure, something that is to come, although it will never fully arrive [DER 05]. Furthermore, Derrida argues that "the other is what is never inventable and will never have waited for your invention. The call of the other is a call to come, and that happens only in multiple voices" [DER 07]. To find a way out of the present impasse towards L´avenir, Derrida proposed practicing deconstruction, which means not being enclosed or dominated by the same or majority [WHI 07].

SOEDWAM is introducing L´avenir II

However, a philosophical and cybernetical perspective must be adopted to decipher the foundations of integrative models of serial entrepreneurship enhancing entrepreneurially driven marketing strategies. Therefore, the Design Weltanschauung model by [KAM 21] needs to be examined via the SOEDWAM, as shown in Figure 10.17.

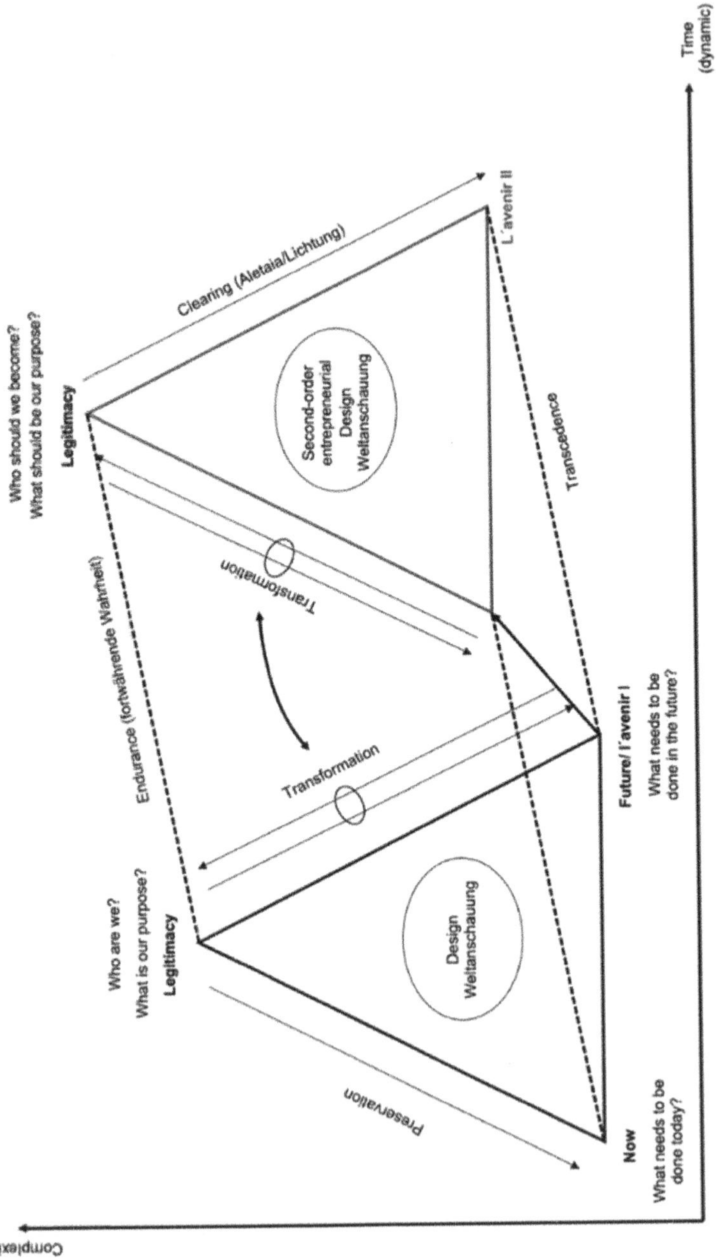

Figure 10.17. *SOEDWAM introducing L´avenir II*

For "usual" companies, the Design Weltanschauung model by [KAM 21] remains the valid model in Design Science to operate successfully in cyber-physical realities. However, the findings in previous chapters require developing an additional layer to enhance entrepreneurially driven marketing strategies. Therefore, a philosophical and cybernetical dimension must be implemented to introduce L´avenir II, enabling marketers to handle even higher complexity and dynamic issues. Therefore, the marketer needs to follow the fortwährenden Wahrheit (Endurance) to maintain its raison d´etre following the analogy of L´avenir II.

Simultaneously, marketers must be able to transcend current knowledge and use Dasein to base their operations on SOEDWAM. They need to operate from a constructivist perspective.

Knowledge does not represent "objective" facts, but is a compendium of concepts, conceptual relationships and rules that have proven helpful in the world [FOE 84]. Foerster introduced the concept of second-order cybernetics and got rid of the control and regulation euphoria that characterized the first-order cybernetics of the time.

In second-order cybernetics, the observer and the observed are inextricably intertwined. Second-order cybernetics brings into play the constant obligation to reflect the idiosyncrasies and blind spots of the observer, never to separate assertions from the individual who makes them, but always to evaluate them in a serious sense as the product of an individual.

Going a step further, radical constructivism can be seen as an epistemology that offers a pragmatic approach to questions of reality, truth and human understanding [GLA 84]. Knowledge is categorized according to its realizability in the realm of experience rather than according to the traditional philosophical position that it is constitutive of truth, i.e. that it corresponds to objective reality (phenomenology).

The two basic principles of radical constructivism are as follows:

– Knowledge is not passively received through the senses, but actively constructed by the cognizing subject - in this context, the learner.

– The function of cognition is the organization of the world of experience rather than the discovery of an independent reality [WAL 20].

Now, what about the light of truth covered by the right side of the triangle? Is the clearing also, in this case, a more original dimension in which there can only be something like truth? Yes, and even in an even deeper sense, because the essence of truth as αλήθεια, the coming forth from concealment into unconcealment – i.e. a coming to light that is essentially related to darkness – is the essence of the clearing itself. Heidegger explains the Greek word αλήθεια as a "privative expression." [HEI 79].

On the one hand, the openness of Dasein is being enlightened, and, on the other hand, the unconcealment of Being is becoming enlightened, expressing the original essence of truth, which remains veiled in the word "truth." Heidegger also calls the development of Dasein and the discovery of Being "openness" or "revelation" [HEI 76]. The openness of Dasein, as letting in on the being, is, according to its essence, a "letting-be" [HEI 76]. Thus, it is not only an ontic, but also an ontological relation: as letting-be, letting-be does not only refer to the being, but also only to be, i.e. to "the open and its opennesses, which every being brings with itself, as it were. This open has been conceived by western thought in its beginnings as τα άληθέα, the unconcealed" [HEI 76], in which only something can open: this area of reference is the clearing of being. "The name of this clearing is αλήθεια" [HEI 76].

"Clearing," even more than "truth," is the appropriate "translation" of αλήθεια as the realm of unconcealment arising from concealment. In order to translate the word αλήθεια in its fullest sense as "clearing," the following conditions must be met. If the clearing is the place, the realm of the truth event as a present happening, it must not be thought of as a place already given: rather, it retains the literal sense of "clearing," i.e. it is the breaking open of an opening. Connected with this is the essential relationship of darkness and light (of concealment and unconcealment), for the clearing points to an emergence from the irreducible, harboring darkness into the light [AMO 83]. All the previously described philosophical and cybernetical dimensions are foundations of integrative models of serial entrepreneurship enhancing entrepreneurially driven marketing strategies.

Based on the findings, L´avenir II can be defined as follows:

I) Entrepreneurs need the ability to endure to follow the analogy of l´avenir II. The if-question is replaced by the when-question.

II) Knowing the science is a necessary prerequisite **for participation in** knowledge-exchange on *Augenhöhe*.

III) Risk-taking and focusing on entrepreneurial opportunities instead of boundaries and doubts.

IV) Being creative is a key pillar for entrepreneurs applying l´avenir II.

V) Co-creation is required for entrepreneurs to follow the analogy of l´avenir II.

VI) Referring to Kant, entrepreneurs need to follow the epistomogical "a priori" (before experience) instead of "posteriori" (after experience). This means, that **basing activities solely on experience does not follow the analogy of l'avenir II for serial entrepreneurship in cyber-physical realities. To give an example, Musk developed reusable space vehicles without any experience due to the novelty of his visions, risk-taking, and ability to endure**

Figure 10.18. *Key pillars L´avenir II*

For example, the foundation of Zip2 in 1995, Elon Musk's first company, followed the analogy of L′avenir II. Nowadays, many revenue models of contemporary location-based services, smart cities and web maps are based on Musk's company, founded over 20 years before competitors based their operations on this model. Before the foundation of Zip2, the Dasein of such opportunities was not given to represent the unforeseeable future called L′avenir II.

10.10.4. *Conclusion*

The author concludes by designing a foundational model derived from the data observed to serve as the meta-framework for establishing effective marketing management models in contemporary disruption and change.

For marketers of today and the future, the SOEDWAM would deliver a boost in their capabilities and phenomenologically creative work of creating value. By enhancing the boundaries of serial entrepreneurship and embedding a philosophical and cybernetical dimension, the essence of a competitive advantage can be achieved.

To summarize, entrepreneurs must ask themselves three key questions, building upon the SOEDWAM:

– Who should the author become, and what should our purpose be?

– What needs to be done in the future?

– How do we transcend the future?

Figure 10.19 displays the foundations of the SOEDWAM, where entrepreneurially driven marketers can navigate the turbulent markets within the dimensions of "transforming" the organizational raison d′etre by the structural dynamics of an ambidextrous embodiment (adapted by [KAM 21]). While focusing on the dimension of clearing/Lichtung, it is essential to follow the analogy of L′avenir II. Therefore, it illustrates how designer-based serial entrepreneurship can be applied because the cyber-physical markets are moving towards increasing complexity and dynamics, leading to a reduced time to respond effectively and innovate.

Special attention must be given to the philosophical foundation, and idealism, pragmatism and phenomenology must be detailed here.

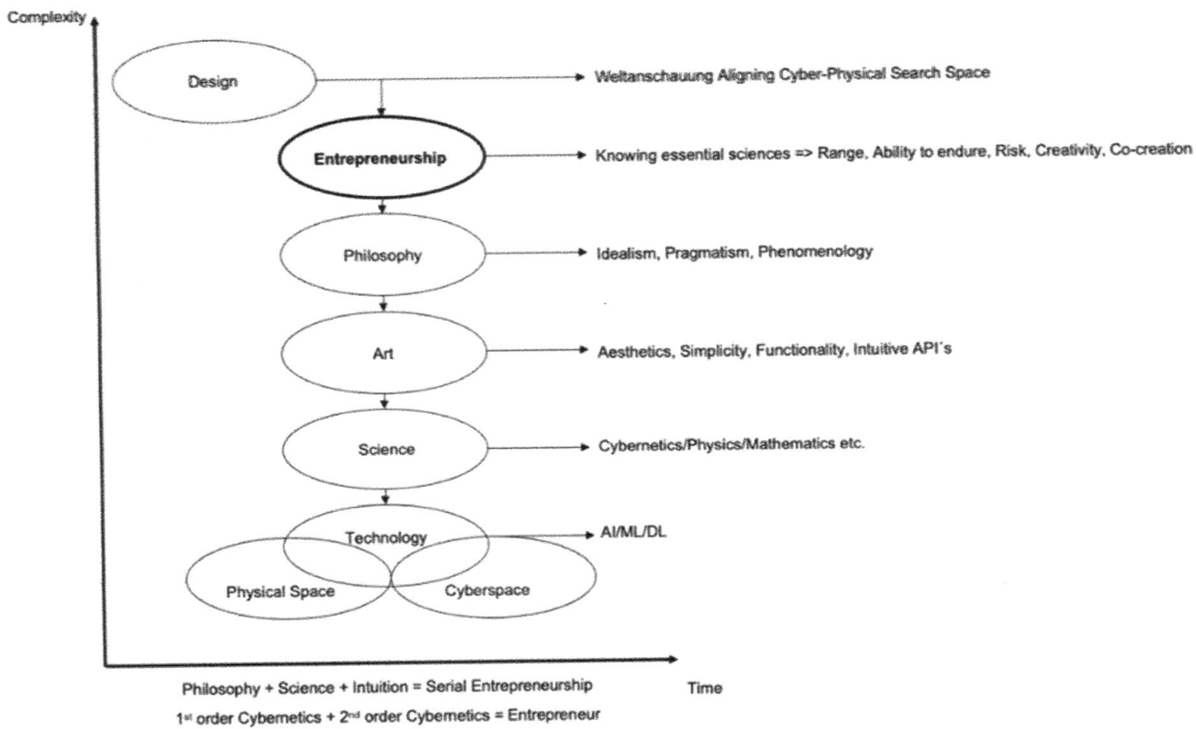

Figure 10.19. *Describing the SOEDWAM applied and aligned to diverse essential fields for serial entrepreneurship enhancing entrepreneurially driven marketing*

Scan this QR code to access high-resolution and interactive representations of Figures 10.9 to 10.15 from this chapter

10.10.5. *References*

[ARU 10] ARUN R., SURESH V., VENI MADHAVAN C.E. et al., "On finding the natural number of topics with latent Dirichlet allocation: Some observations", in ZAKI M.J., YU J.X., RAVINDRAN B. et al. (eds), *Advances in Knowledge Discovery and Data Mining, Part I. 14th Pacific-Asia Conference, PAKDD 2010, Hyderabad, India, 21–24 June 2010, Proceedings.*, Springer, New York, pp. 391–402, doi: 10.1007/978-3-642-13657-3_43, 2010.

[BEE 93] BEER S., *Designing Freedom*, House of Anansi, Toronto, 1993.

[DER 05a] DERRIDA J., *Paper Machine*, Stanford University Press, Stanford, CA, 2005.

[DER 05b] DERRIDA J., *Rogues: Two Essays on Reason*, Stanford University Press, Stanford, CA, 2005.

[DER 07] DERRIDA J., *Psyche: Inventions of the Other*, Stanford University Press, Stanford, CA, 2007.

[FRE 22] FRENCH M.F., Imminence over immanence: A lesson in L'Avenir, Patheos Blog, available at: https://www.patheos.com/blogs/mariafrancescafrench/2022/03/imminence-over-immanence-a-lesson-in-lavenir/ [Accessed 29 March 2022], 2022.

[HEI 72] HEIDEGGER M., *Sein und Zeit*, 12th edition, Max Niemeyer, Tübingen, 1972.

[HEI 76] HEIDEGGER M., *Vom Wesen der Wahrheit*, 6th edition, Vittorio Klostermann, Frankfurt am Main, 1976.

[KAM 21a] KAMRAN Q., Managing complexity in marketing: from a design Weltanschauung, Thesis, University of Twente, Enschede, 2021.

[KAM 21b] KAMRAN Q., TOPP S., BECKER M., "The structure and evolution of the marketing field: A content analysis of five decades of research within the Academy of Marketing Science Journals", *Journal of the Academy of Marketing Science*, vol. 49, pp. 1021–1047, doi: 10.1007/s11747-020-00758-x, 2021.

[KAN 81] KANT I., *Kritik der reinen Vernunft*, Johann Friedrich Hartknoch, Riga, 1781.

[MAR 07] MARTINON J., *On Futurity: Malabou, Nancy, and Derrida*, Palgrave Macmillan, Basingstoke, 2007.

[MIN 20] MINTZBERG H., AHLSTRAND B.W., LAMPEL J., *Strategy Safari: A Guided Tour Through the Wilds of Strategic Management*, 2nd edition, FT Publishing International, Halow, 2020.

[NEW 09] NEWMAN D., ASUNCION A., SMYTH P. et al., "Distributed algorithms for topic models", *Journal of Machine Learning Research*, vol. 10, pp. 1801–1828, available at: https://www.jmlr.org/chapters/v10/newman09a.html, 2009.

[VON 84a] VON FOERSTER H., "On cybernetics of cybernetics and social theory", in ROTH G., SCHWEGLER H. (eds), *Self-Organizing Systems*, Campus Verlag, Frankfurt am Main, pp. 102–105, 1984.

[VON 84b] VON GLASERSFELD E., "An introduction to radical constructivism", in WATZLAWICK P. (ed.), *The Invented Reality: How Do We Know What We Believe We Know?*, Norton, New York, 1984.

[WAL 20] WALSHE G., "Radical constructivism—von Glasersfeld", in AKPAN B., KENNEDY T. (eds), *Science Education in Theory and Practice*, Springer, Cham, pp. 359–371, doi: 10.1007/978-3-030-43620-9_24, 2020.

[WHI 07] WHITE E., "A passage toward the other: The legacy of Jacques Derrida (1930–2004)", *European Legacy*, vol. 12, no. 4, pp. 407–408, doi: 10.1080/10848770701395926, 2007.

[YAO 09] YAO L., MIMNO D., MCCALLUM A., "Efficient methods for topic model inference on streaming document collections", in *Proceedings of the 15th ACM SIGKDD International Conference on Knowledge Discovery and Data Mining*, Association for Computing Machinery, New York, pp. 937–946, doi: 10.1145/1557019.1557121, 2009.

11

The Use of AI in Human Resource Management: Barriers, Opportunities and Trends

The advancement of artificial intelligence (AI) has driven significant changes in Human Resource Management (HRM), affecting practices such as recruitment, training and performance evaluation. This chapter aims to understand whether AI is being applied in HRM and how it is being implemented while also examining the emerging ethical and legal challenges and future trends in its use.

The objectives of this chapter are (i) to identify the main reasons for the (non-) adoption of AI in HRM, including technical, financial or cultural barriers; (ii) to investigate the level of knowledge and perception of HR managers regarding the benefits and limitations of AI; (iii) to explore companies' future intentions regarding AI adoption in HRM processes; and (iv) to analyze the ethical and legal challenges associated with the implementation of AI in people management practices.

The adopted methodology is qualitative, based on semi-structured interviews with 16 HR managers from companies of different sectors and sizes. The results indicate that AI has contributed to optimizing processes, improving decision-making and personalizing professional development experiences. However, challenges such as algorithmic bias, transparency in automated decisions and employee data security have also been identified.

It is concluded that, although AI brings substantial benefits to HRM, its implementation requires strategies that ensure fairness, ethics and trust in the systems used.

In the future, an increased integration of AI into HRM processes is expected, making continuous training of professionals in the field essential for effective and responsible adoption.

Chapter written by Pedro Miguel Torres BARROS and Carolina MACHADO.

11.1. Introduction

Artificial intelligence (AI) has been significantly transforming Human Resource Management (HRM), introducing new dynamics in processes such as recruitment, training, performance evaluation, talent management and benefits management [TAM 19, EKU 23]. With advanced data processing and analysis capabilities, AI promises to improve the efficiency and effectiveness of these functions. However, its implementation raises complex ethical and organizational issues, such as algorithmic transparency, bias in decision-making processes and the privacy of employee data.

Recent studies indicate that AI allows the identification of patterns that support strategic decision-making [LIU 18], and it is essential to ensure its use in an equitable and transparent manner [DWI 21]. In this context, this study examines the application of AI in HRM across various areas, including recruitment and selection, training and development, performance evaluation, talent management and benefits management. In addition, the ethical and legal challenges of AI implementation are analyzed, as well as future trends and the skills needed for its effective integration into HRM.

The choice of this problem arises from the need to investigate, given the rapid and constant advancement of AI, how it can contribute to HRM and what problems may arise in its use.

The initial question supporting this research is:

"Why are companies (or are they not) using AI in their HRM processes, and what are the perceptions, barriers and intentions associated with its future use?"

The overall objective was defined as:

"To understand the reasons why companies use (or do not use) AI in HRM processes and to explore perceptions about its benefits, barriers and potential future applications."

Specific objectives were also defined, namely:

– to identify the main reasons that lead (or do not lead) to the use of AI in HRM, including technical, financial or cultural barriers;

– to investigate the level of knowledge and perception of HR managers about the benefits and limitations of AI;

– to explore the future intentions of companies regarding the adoption of AI in HRM processes;

– to analyze the ethical and legal challenges associated with the implementation of AI in people management practices.

The investigation employs a qualitative approach, conducted through semi-structured interviews with 16 Human Resources (HR) managers from companies across various sectors and sizes. This method allows us to capture perceptions and practical experiences, contributing to a deeper understanding of the impact of AI on HRM. The analysis focuses on the ethical and effective integration of AI, proposing ways to maximize its benefits while simultaneously mitigating the associated risks. In addition, it aims to understand the managers' level of knowledge about AI, the challenges they encounter in its adoption and their future intentions for its implementation.

In structural terms, the chapter is organized into four main sections. Section 11.2 corresponds to the literature review, where the HRM processes selected for this work are initially addressed, namely recruitment and selection, training and development, performance evaluation, talent management and benefits management. Within the scope of the literature review, a second sub-section dedicated to AI is presented, exploring its concept, characteristics and applications, as well as the challenges associated with it. This sub-section culminates in a comparison between traditional AI and Generative AI (GenAI). A third sub-section of the literature review examines the application of AI in previously defined areas and processes of HRM, as well as its benefits, challenges and ethical considerations arising from its use.

Section 11.3 is dedicated to the methodology, which includes the research strategy, the data collection method, the definition and characterization of the sample and the method of analysis and discussion of the results.

Section 11.4 covers the analysis and discussion of the results take place, where the perspectives of each HR manager interviewed are presented, according to a previously defined script and the proposed objectives. At the end of this analysis, a Good Practices guide for the use of AI in HR management, authored by the author, has also been developed.

Finally, section 11.5 presents the conclusions of the work, based on the defined objectives, and also outlines the limitations of the research, as well as offers suggestions for future research.

11.2. Theoretical framework

This chapter aims to theoretically frame the relationship between HRM and AI, establishing the necessary conceptual foundations for understanding its integration into organizations. Initially, it addresses the evolution and main functions of HRM, with a special focus on the contemporary challenges facing the area. Next, it presents the concept of AI, its main characteristics, applications and limitations. Ultimately, it examines the integration of AI into HRM, highlighting its practical applications, benefits and associated ethical considerations.

11.2.1. *HRM*

HRM currently plays a strategic role in organizations, being responsible for aligning organizational objectives with the needs of employees. Its evolution reflects economic, technological and social transformations, which have contributed to the transition from a predominantly administrative function to a central area in creating competitive advantage [KAU 15].

11.2.1.1. *A brief evolution of HRM*

Modern HRM emerged in the 1980s with the appreciation of human capital as a strategic asset. Works such as those by [FOM 84] and [BEE 84] introduced the idea of alignment between HRM practices and organizational strategy, promoting high-performance work systems based on training, compensation and performance evaluation [KAU 15].

Subsequently, [WRI 02] advocated an integrated approach between micro and macro levels of HRM, allowing for a systemic analysis of practices. More recently, HRM has consolidated itself as a strategic function, driven by digitalization and the introduction of technologies such as big data and AI, which have expanded the capacity for analysis and decision support, although they raise relevant ethical challenges [KAU 15, COR 22].

11.2.1.2. *Main functions of HRM*

HR functions include recruitment and selection, training and development, talent management, performance evaluation and benefits management. These areas directly contribute to organizational performance and have been significantly influenced by technological and organizational transformations.

11.2.1.2.1. Recruitment and selection

Recruitment and selection have evolved from operational processes to strategic functions, in a context marked by talent shortages and increasing workforce diversity [DER 16]. Organizations are resorting to increasingly sophisticated methods to attract and select candidates aligned with their needs and culture.

Despite the ongoing validation of traditional methods, such as structured interviews and psychometric tests, challenges arise related to fairness, candidate experience and the integration of new technologies [POT 21]. The use of digital and AI tools raises ethical and methodological concerns, requiring caution in their implementation [DER 16, POT 21, GOM 22].

11.2.1.2.2. Training and development

Training and development are essential for organizational competitiveness, especially in contexts of continuous change [MAB 03]. These practices must strike a balance between organizational objectives and individual needs, promoting continuous learning and enhancing employee well-being.

The success of training actions depends on management involvement, alignment with development plans and the existence of pre- and post-training activities that ensure the effective transfer of skills to the work context [SAN 03]. Digitalization and remote work have reinforced the need to continuously evaluate the effectiveness of training and maintain employee motivation [URB 21].

11.2.1.2.3. Talent management

Talent management is now a strategic priority, influenced by globalization, skills shortages and technological advances [MEY 13]. While more traditional approaches view talent as an innate and exclusive characteristic [REI 08], more recent perspectives advocate an inclusive vision, based on the continuous development of human potential [MEY 13].

Attracting, developing and retaining talent requires flexible practices adapted to global contexts, integrating diversity, mobility and social responsibility [FAR 10, TAR 10, CLA 19].

11.2.1.2.4. Performance appraisal

Performance appraisal is a central tool in HR, supporting decisions related to rewards, promotions and development [ROB 03]. Participative approaches, based on continuous feedback and a clear definition of objectives, increase the acceptance and effectiveness of these systems.

More recently, the use of HR analytics has contributed to reducing subjectivity and improving the perception of fairness by using objective data in decision-making support [SHA 17]. These tools complement, but do not replace, the relational dimension of the evaluation process [CAP 17].

11.2.1.2.5. Benefits management

Benefits management plays a significant role in employee satisfaction, motivation and retention. Clear and effective communication is essential to ensure that employees understand and value the benefits available [FRE 04].

Offering benefits aligned with physical, psychological and social well-being contributes to strengthening organizational bonds and enhancing company competitiveness [DUL 09, GUE 17].

11.2.1.3. *Contemporary challenges in HRM*

HRM currently faces complex challenges associated with economic transition, pressure for efficiency, sustainability and the integration of new technologies [THO 11, STO 15].

11.2.1.3.1. Economic transition and the strategic role of HRM

The knowledge-based economy requires highly skilled employees and HR practices that foster autonomy, innovation and continuous learning [STO 15]. However, pressure for financial results can limit investment in human capital, creating tensions between economic objectives and employee development [THO 11].

The integration of AI reinforces the strategic role of HR, requiring new skills and a redefinition of traditional practices [ARS 22].

11.2.1.3.2. Paradoxes and tensions in HR functions

HRM faces paradoxes such as the need to promote innovation while maintaining organizational trust. In hybrid and technologically advanced work contexts, managing trust between humans and AI systems is now also fundamental [ARS 22, SOD 06].

11.2.1.3.3. Sustainability, social responsibility and innovation in HR

Sustainability in HRM goes beyond the environmental dimension, integrating concerns about well-being, equality and social impact [POD 20]. Technological innovation, including AI, can contribute to more efficient and humane practices, provided it is implemented ethically and responsibly [ARS 22, THO 11].

11.2.2. *AI*

11.2.2.1. *Fundamental concepts and evolution of AI*

AI is an interdisciplinary field that aims to develop systems capable of performing tasks that would normally require human intelligence, such as learning, reasoning and decision-making [LIU 18]. Its development has been driven by advances in computing power, the availability of large datasets and the evolution of machine learning algorithms.

[KAP 19] defined AI as the ability of systems to interpret data, learn from it and adaptively use that knowledge to achieve specific goals. This definition highlights flexibility as a central characteristic of AI, distinguishing it from traditional computing systems.

[WAN 19] adds that AI is characterized by its ability to operate in dynamic environments, dealing with uncertainty, incomplete information and limited resources. This adaptability makes AI particularly relevant in complex organizational contexts.

11.2.2.1.1. Definition

AI can be understood as a set of technologies that simulate human cognitive abilities, using advanced algorithms for pattern recognition, learning and problem-solving [LIU 18, MEH 23].

These systems are capable of processing large volumes of data with high speed and accuracy, thereby supporting complex decision-making processes.

[KAP 19] distinguished between different levels of AI, ranging from restricted AI, which focuses on specific tasks, to more advanced and theoretical forms, such as general AI and superintelligence. This classification enables us to understand the current stage of technology development and its limitations.

11.2.2.1.2. Characteristics of AI

Among the main characteristics of AI, the ability to learn continuously, autonomy and operational flexibility stand out [WAN 19]. These attributes allow systems to adjust their performance as they receive new data.

[MEH 23] also highlighted AI's ability to identify complex patterns and support real-time decision-making, overcoming human cognitive limitations. However, these advantages require high levels of reliability and transparency.

[SAE 23] argued that attributes such as robustness, resilience and explainability are essential to ensure the safe and ethical use of AI. Without these characteristics, systems can generate distrust and organizational risks.

11.2.2.1.3. Advances and applications of AI: impacts on society and businesses

Advances in machine learning and deep learning have enabled AI to achieve high levels of performance in areas such as image recognition, natural language processing and predictive analytics [WAN 19, MOR 22]. These technologies have been widely applied in various sectors, including healthcare, finance, industry and services.

In the business context, AI contributes to process automation, service personalization and improved operational efficiency [HAE 19, DWI 21]. Simultaneously, it plays a significant role in addressing global challenges, such as the sustainable development goals, by supporting initiatives in education, health and natural resource management [GÓR 20, TOM 20].

Despite the benefits, the application of AI requires adequate infrastructure and responsible governance policies; otherwise, it risks deepening social and economic inequalities [DWI 21].

11.2.2.2. *Challenges and limitations of AI*

The implementation of AI faces significant challenges, with algorithmic bias being one of the most relevant. When trained with biased data, systems can reproduce existing discriminations, affecting processes such as recruitment and evaluation [LIU 18, SAE 23].

Another central challenge is the lack of transparency of complex models, often described as "black boxes", which makes it difficult to explain decisions and compromises trust in the systems [DWI 21]. Explainability and auditability thus become fundamental requirements [KUS 20, SAE 23].

Ethical and privacy issues are also highly relevant, given the intensive use of personal data [LIU 18]. In addition, automation can have social implications, such as job displacement, necessitating the development of professional retraining strategies [KAP 19].

11.2.2.3. *Traditional AI versus generative AI*

Traditional AI relies on the analysis of structured data and the identification of patterns to support specific decisions, being widely used in tasks such as classification and prediction [MAR 23]. These systems have high accuracy but limited creative capacity.

In contrast, generative AI is capable of creating new content, such as text, images or music, using advanced models, including generative and transforming neural networks [MAR 23, WOH 24]. This approach stands out for its flexibility and creative capacity but raises greater challenges in terms of control, predictability and bias.

Despite the growth of generative AI, traditional AI remains central in more regulated contexts, where reliability and interpretability are essential [MAR 23].

11.2.3. *Integrating AI into HRM*

11.2.3.1. *Applications of AI in HRM processes*

AI has been transforming HR processes, enabling greater efficiency, personalization and decision support [TAM 19]. Its application covers areas such as recruitment, training, performance evaluation and talent management.

11.2.3.1.1. Recruitment and selection

In recruitment and selection, AI enables the automation of resume screening, the identification of success patterns and the reduction of operational costs [TAM 19, QAM 21]. Machine learning-based tools analyze large volumes of data, increasing the speed and consistency of decisions.

Digital interviews and automated profile analysis contribute to greater objectivity, although they raise concerns related to dehumanization and algorithmic bias [BUD 22, PAN 22]. Effective implementation requires transparency and human oversight.

11.2.3.1.2. Training and development

AI enables the personalization of training programs, adapting content to the individual needs of employees [EKU 23]. Intelligent systems enable the monitoring of progress and the recommendation of learning paths, thereby increasing training effectiveness.

However, excessive reliance on technology can compromise motivation and human interaction, requiring a balance between digital solutions and organizational support [URB 21].

11.2.3.1.3. Performance appraisal

In performance evaluation, AI facilitates the objective analysis of indicators, thereby reducing subjectivity and enabling more informed decisions [SHA 17]. The use of HR analytics contributes to a greater perception of fairness, provided it is complemented by human feedback [CAP 17].

11.2.3.1.4. Talent management

AI facilitates talent identification, forecasting future needs and succession planning [MEY 13]. Predictive analytics enables the alignment of individual skills with strategic objectives, promoting more proactive management.

11.2.3.1.5. Benefits management

In benefits management, AI supports the personalization of packages and effective communication, increasing employee satisfaction and well-being [DUL 09, GUE 17].

11.2.3.2. *Benefits, challenges and ethical considerations of AI in HRM*

The main benefits of AI in HR include increased efficiency, improved decision-making, personalized experiences and reduced costs [TAM 19]. These advantages reinforce the strategic role of the HR function.

Despite the benefits, the use of AI in HRM raises ethical challenges related to bias, privacy, transparency and trust [DWI 21]. Responsible adoption requires clear policies, ongoing training and human oversight.

11.3. Methodology

This chapter adopts a qualitative approach, considered appropriate for a deeper understanding of the perceptions, experiences and interpretations of HR managers regarding the use of AI in their organizational contexts. The choice of this approach is justified by the exploratory nature of the phenomenon under study, since the adoption of AI in HRM is still in an initial and consolidation phase, requiring an interpretative analysis of the meanings attributed by the participants.

The research strategy employed was exploratory, allowing for the analysis of an emerging phenomenon without imposing rigid hypotheses from the outset and prioritizing an understanding of the phenomenon from the participants' perspective. This methodological option aligns with the study's objective, which is to understand not only whether AI is used but also how it is perceived, interpreted and integrated into HRM practices.

11.3.1. *Sample selection and participants*

The study sample consists of 16 HR managers, selected based on their role and direct involvement in HRM processes. This selection criterion ensures that participants possess relevant knowledge and practical experience in the topics covered, specifically the adoption of digital technologies and the potential application of AI.

The organizations represented belong to different sectors of activity and vary in size, allowing us to capture a diverse range of organizational contexts. This diversity contributes to a more comprehensive understanding of the phenomenon, without claims of statistical generalization, but reinforces the interpretative validity of the results.

11.3.2. *Data collection*

Data collection was conducted through semi-structured interviews, which are considered suitable for exploring complex topics and allowing flexibility in the interview process. The interview guide was developed based on a literature review and the research objectives, addressing topics such as the degree of AI adoption, areas of application, benefits, barriers, ethical issues and future perspectives.

The interviews were conducted individually, ensuring the anonymity and confidentiality of the participants, in order to promote an environment of trust and encourage open and honest responses.

11.3.3. *Data analysis*

The data collected were analyzed through content analysis, which allowed for the identification of thematic categories and subcategories aligned with the study's objectives. The analysis process involved a complete reading of the interviews, coding of the units of meaning and subsequent organization of the data into analytical categories.

This procedure ensured coherence between the empirical data and the theoretical framework, as well as rigor in the interpretation of the results.

11.3.4. *Sample characterization*

The 16 companies (E1 to E16), indicating the sector of activity according to the CAE (Portuguese Classification of Economic Activities) and the company size:

– E1: Section N: administrative and support service activities; large size.

– E2: Section E: water collection, treatment and distribution; sanitation, waste management and pollution control; large size.

– E3: Section M: consulting, scientific, technical and similar activities; small size.

– E4: Section J: information and communication activities; large size.

– E5: Section O: public administration and defense; mandatory social security; large size.

– E6: Section C: manufacturing industries; large size.

– E7: Section N: administrative and support service activities; medium size.

– E8: Section N: administrative and support service activities; large size.

– E9: Section K: financial and insurance activities; medium size.

– E10: Section J: information and communication activities; large size.

– E11: Section C: manufacturing industries; small size.

– E12: Section N: administrative and support service activities; medium size.

– E13: Section J: information and communication activities; large size.

– E14: Section M: consulting, scientific, technical and similar activities; large size.

– E15: Section G: wholesale and retail trade; repair of motor vehicles and motorcycles; medium size.

– E16: Section E: water collection, treatment and distribution; sanitation, waste management and pollution control; medium size.

The 16 participants (PE1 to PE16), with information on gender, age, academic background, current position, time in the organization and time in the role:

– PE1: female, 48 years old, bachelor's degree, Branch Director, seven years in the organization and seven years in the role.

– PE2: female, 22 years old, bachelor's degree, Senior Human Resources Technician, one year and five months in the organization and one year and five months in the role.

– PE3: female, 50 years old, master's degree, CEO, twelve years in the organization and twelve years in the role.

– PE4: female, 25 years old, master's degree, Recruiter, ten months in the organization and ten months in the role.

– PE5: male, 41 years old, bachelor's degree, Head of Human Resources Section, 24 years in the organization and 24 years in the role.

– PE6: male, 31 years old, master's degree, Recruiter, two years and six months in the organization and two years and six months in the role.

– PE7: female, 27 years old, postgraduate degree, Talent Acquisition Manager, one year and two months in the organization and one year and two months in the role.

– PE8: female, 25 years old, postgraduate degree, Talent Consultant, four months in the organization and four months in the role.

– PE9: male, 28 years old, master's degree, Recruiter, two years and eleven months in the organization and two years and eleven months in the role.

– PE10: female, 24 years old, postgraduate degree, Head of Employee Secondment and Mobility, eight months in the organization and eight months in the role.

– PE11: female, 22 years old, postgraduate degree, Human Resources and Quality Manager, one year and one month in the organization and one year and one month in the role.

– PE12: female, 32 years old, bachelor's degree, Tech Recruiter, three months in the organization and three months in the role.

– PE13: female, 30 years old, master's degree, Senior Human Resources Technician, four years in the organization and four years in the role.

– PE14: female, 31 years old, bachelor's degree, Recruiter Analyst, three years in the organization and three years in the role.

– PE15: female, 28 years old, master's degree, People Operations Specialist, three years and five months in the organization and three years and five months in the role.

– PE16: female, 35 years old, postgraduate degree, Senior Human Resources Technician, six years in the organization and two years in the role.

11.4. Analysis and discussion of results

This section aims to analyze and discuss the results obtained from interviews conducted with HR managers, seeking to understand how AI is being integrated into HRM processes, as well as the perceptions, benefits, barriers and challenges associated with its use. The analysis of the results is conducted in conjunction with the theoretical framework presented in previous sections, enabling the establishment of relationships between the empirical evidence collected and the existing literature.

Thus, the section is organized around the adoption of AI in HRM, managers' perceptions regarding its use, areas of application and non-application of AI and finally, future perspectives regarding its integration into HRM processes.

11.4.1. *Adoption of AI in HRM*

Regarding the adoption of AI in HRM, the results reveal that its use is not yet widespread among the organizations analyzed. Of the 16 companies included in the study, only seven report currently using AI-based tools in their HR processes, demonstrating that the adoption of this technology is still in an initial and uneven phase.

The companies that already use AI primarily rely on solutions that automate administrative and internal communication tasks, as well as provide support in recruitment and selection processes.

In particular, the use of generative AI tools stands out, as they are used to support candidate evaluation, prepare administrative documentation and respond to internal questions. One of the interviewees explicitly mentions this practice by stating: "Yes, we use ChatGPT prompts, 'an internal ChatGPT', which allows us to evaluate a candidate" (PE6).

This testimony demonstrates that AI is primarily used as a decision support tool, assisting HR professionals in organizing and analyzing information, while still relying on human judgment. This approach aligns with the literature, which argues that the adoption of AI in HRM tends to be gradual and complementary, especially in organizational contexts where reservations persist about the total automation of processes [TAM 19, PAN 22].

Regarding companies that have not yet adopted AI, the primary reasons cited are a lack of resources, the absence of internal skills and concerns about the impact of technology on people management processes. These factors contribute to a more cautious stance, reflecting the need for a careful assessment of the risks and benefits associated with implementing AI.

11.4.2. *Perceptions about AI in HRM*

In general, HR managers demonstrate a mostly positive perception regarding the potential of AI in HRM. AI is viewed as a tool that can enhance process efficiency, reduce the time spent on repetitive tasks and facilitate strategic decision-making.

However, this positive perception is accompanied by a cautious stance, with the importance of human oversight and the responsible use of technology being frequently emphasized. Managers recognize that, despite the benefits, AI raises significant challenges that must be carefully considered.

11.4.2.1. *Perceived benefits*

Among the primary benefits of using AI in HR, respondents emphasize improved operational efficiency, process optimization and enhanced support for decision-making. The automation of administrative tasks frees up time for HR professionals to focus on activities of greater strategic value, such as talent development and managing interpersonal relationships.

The results also indicate that AI can contribute to greater accuracy in data analysis, allowing for more informed and well-founded decisions. This perception aligns with the literature, which identifies AI as a facilitator of data-driven decision-making, particularly in complex organizational contexts [SHA 17, EKU 23].

11.4.2.2. *Barriers to the use of AI in HRM*

Despite the identified benefits, HR managers recognize several barriers to the adoption of AI. Among the primary challenges are resistance to change, a lack of internal technical expertise, and the associated costs of implementing the technology.

Resistance to change is associated with the fear of losing control over decision-making processes and distrust of automated decisions. One of the interviewees underlines the importance of a gradual implementation of the technology, stating the need for "a gradual implementation strategy, raising awareness and training people and teams, ensuring transparency in the decision-making process" (PE5).

This testimony demonstrates that the acceptance of AI strongly depends on how it is introduced into the organization, reinforcing the importance of communication, training and employee involvement in the change process, aspects also highlighted in the literature on digital transformation and technology adoption [SUE 19, PAN 22].

11.4.2.3. *Essential skills for using AI in HR*

The results indicate that the effective use of AI in HRM requires the development of new skills among HR professionals.

In addition to the technical skills associated with using the tools, managers emphasize the importance of analytical, interpretive and ethical skills.

The ability to understand how AI systems work and to correctly interpret the results generated is seen as essential to ensure fair and well-founded decisions. This need for continuous training emerges as a central element for the responsible adoption of technology, as studies have underscored the role of analytical and ethical skills in the use of AI in HRM [EKU 23].

11.4.2.4. *Ethical and legal considerations*

Ethical and legal issues associated with the use of AI in HR emerge across the discourse of the interviewees. Among the main concerns are algorithmic bias, the transparency of automated decisions and the protection of employees' personal data.

Managers demonstrate awareness of the risks associated with the use of automated systems and emphasize the importance of ensuring that decisions made with the support of AI are understandable, auditable and aligned with principles of fairness and justice, in line with the ethical concerns widely discussed in the literature on AI [LIU 18, DWI 21].

11.4.2.5. *How to overcome ethical and legal challenges*

To mitigate the ethical and legal challenges associated with AI, respondents emphasize the need for continuous human oversight and clear policies regulating the use of the technology. Transparency and communication are key factors in increasing trust in AI systems.

One participant explicitly mentions the importance of "human communication and oversight of actions performed by machines" (PE16) to avoid problems and strengthen trust in the technology.

This perspective reinforces the idea that AI should complement human work, not replace it, as advocated by several authors in the context of integrating AI into HRM [ARS 22].

11.4.3. *Areas of application and/or non-application of AI in HRM*

Analysis of the results allows us to identify areas of HR where AI is considered most useful, as well as areas where its application is viewed with greater reservation. This distinction highlights the need to strike a balance between the efficiency provided by technology and the importance of the human factor in people management.

11.4.3.1. *Areas where AI may have limitations*

Respondents identify several areas of HR where the use of AI should be approached with caution, especially in processes that require empathy, sensitivity and human judgment. Recruitment interviews are often cited as an example of a process that should maintain a strong human element.

One participant states that "it's important that it's done by a human, you can talk about different topics, the conversation evolves differently, from candidate to candidate, from position to position" (PE1). Another interviewee adds that "a recruiter has a different analysis, a more emotional analysis" (PE7), reinforcing the importance of human interaction.

In addition, PE13 emphasizes that human contact is essential to truly get to know candidates, stating that "we don't know through codes (...) only contact with people allows us to understand who is on the other side".

This evidence demonstrates that, although AI can support initial stages of recruitment, the final decision continues to depend heavily on human intervention, corroborating the literature that advocates a hybrid approach between technology and human judgment in selection processes [DER 16, TAM 19].

11.4.3.2. *Areas where AI can be most useful*

On the contrary, managers recognize that AI can be particularly useful in analytical and repetitive tasks, such as screening resumes, analyzing performance data and monitoring talent management indicators. In these contexts, the technology is viewed as a valuable support that enhances the efficiency and consistency of processes.

11.4.4. *Future perspectives of AI in HRM*

Regarding future prospects, HR managers anticipate a gradual increase in the use of AI in HRM processes, albeit cautiously and progressively. Most respondents believe that future adoption will depend on the development of internal skills, the evolution of regulations and the ability of organizations to ensure ethical and transparent use of the technology, an aspect also highlighted in recent studies on the future of AI in HRM [DWI 21, ARS 22].

In summary, the results of this chapter show that AI is perceived as a tool with high potential for HRM but whose implementation requires a balanced approach that combines technological innovation with human oversight and ethical responsibility.

11.5. Best practice guide for using AI in HRM

This document presents a best practices guide for the use of AI in HRM, developed based on evidence gathered throughout this study, both from a literature review and empirical analysis. The guide aims to provide guidance for the ethical, effective and strategic implementation of AI in HRM processes, considering the identified benefits, the challenges faced, and the ethical concerns expressed by the HR managers interviewed. The goal is to support organizations in the responsible adoption of this technology, maximizing its potential and mitigating associated risks.

11.5.1. *Fundamental principles*

The use of AI in HR should be based on the principle of complementarity, viewed as a support for HR managers' decision-making rather than a substitute for human decision-making. It is essential to ensure that the implementation of AI respects ethical and transparency principles, through the definition of clear guidelines, the explanation of algorithmic decision-making processes to employees and the obtaining of explicit consent for the collection and processing of personal data. In parallel, all critical decisions should maintain human oversight, and regular audits of AI systems are recommended to identify and mitigate potential biases.

11.5.2. *Applications and best practices*

In recruitment and selection, AI can support the screening of resumes based on objective criteria, contributing to the reduction of biases. However, human intervention in interviews and final decisions, as well as transparency in the criteria used, must be ensured.

In training and development, AI can be used to personalize training programs, track employee progress and recommend further training. It is advisable to combine technological tools with human support.

Regarding performance and talent management, AI can assist in identifying productivity patterns and anticipating development and retention needs.

Performance evaluation should follow a hybrid approach that combines data analysis with human feedback.

In administrative and compliance processes, AI can automate repetitive tasks, such as payroll processing, while always ensuring human validation, data security, access control and regular audits to ensure fairness and compliance.

11.5.3. *Challenges and solutions*

One of the main challenges identified relates to the risk of algorithmic bias, making frequent audits and testing of systems with diverse data essential. In addition, it is crucial to maintain interpersonal relationships within the organizational context, striking a balance between automation and human interaction, in order to preserve empathy and the human dimension in the workplace. Resistance to change is also a significant challenge, which should be addressed through awareness-raising and team training, as well as the gradual introduction of AI, accompanied by human supervision.

11.5.4. *Gradual and sustainable implementation*

The implementation of AI in HR should be phased in, beginning with needs analysis and the definition of strategic objectives, followed by pilot projects in specific processes. Subsequently, there should be a progressive expansion of AI use, always ensuring continuous human supervision, monitoring and improvement through audits and user feedback. Collaboration with experts in the fields of ethics, technology and law is also recommended to ensure the responsible and sustainable implementation of these initiatives.

Concluding note

In summary, AI presents high potential for HR, contributing to process optimization and more strategic decision-making. However, its use must be guided by principles of ethics, transparency and human oversight, ensuring that it complements, rather than replaces, human interaction. A balanced implementation is essential to maximize the benefits of the technology while safeguarding the trust and well-being of employees.

11.6. Conclusion

This chapter aimed to analyze the use of AI in HRM, exploring the degree of adoption, the perceptions of HR managers, the associated benefits and barriers, as well as the ethical implications and future perspectives of this technology in organizations. Through a qualitative approach, it was possible to obtain an in-depth understanding of a phenomenon that is still emerging and consolidating in the organizational context.

The results indicate that the adoption of AI in HRM remains limited and uneven. Although some organizations already use AI-based tools, particularly in processes such as recruitment and data analysis, many are still in the initial phase of adoption or choose not to employ these technologies. This reality confirms what has been pointed out in the literature, which identifies digital maturity, available resources and organizational culture as determining factors in the implementation of AI.

In general, HR managers demonstrate a positive perception regarding the potential of AI. Technology is primarily viewed as a tool to support decision-making and optimize processes, thereby reducing the time spent on administrative tasks and enhancing operational efficiency. These results reinforce the perspective defended in the literature that AI should complement, and not replace, the human role in people management.

Despite the identified benefits, the study highlights the existence of significant barriers to the adoption of AI in HRM. High implementation costs, a lack of internal technical expertise and resistance to change are the primary obstacles cited by managers.

These challenges reflect difficulties that have been widely discussed in studies on digital transformation and organizational innovation, underscoring the need for planned change strategies and investment in training.

One of the most relevant aspects of the study concerns the ethical and legal concerns associated with the use of AI in HRM. Issues such as algorithmic bias, the transparency of automated decisions and the protection of personal data were consistently mentioned by participants across the board. These results confirm the importance of responsible AI adoption, supported by clear policies, human oversight and ethical principles, as advocated in the literature.

Regarding application areas, it is concluded that AI is considered more suitable for analytical, repetitive and decision-support tasks, while processes that require empathy, human judgment and direct interaction are deemed less suitable for automation. This distinction reinforces the idea of complementarity between technology and people management, avoiding overly technocentric approaches.

With respect to future prospects, managers anticipate a gradual increase in the use of AI in HRM, albeit cautiously and accompanied by regulation and ethical practices. This view suggests that the integration of AI will be a progressive process, requiring continuous adaptation by organizations and HR professionals.

Limitations of the study and suggestions for future research

Despite the study's contributions, it is important to acknowledge some limitations. The small sample size and the qualitative nature of the research limit the generalizability of the results. In addition, the study reflects managers' perceptions at a specific point in time in a context of rapid technological evolution.

Future research could employ quantitative or mixed methodologies, involving larger samples, as well as comparative analyses of different sectors of activity or organizational contexts. Analyzing the long-term impact of AI and studying specific AI practices in HRM are also relevant areas for future research.

11.7. Declaration

This chapter was edited with the assistance of Grammarly, an AI-powered writing tool, to enhance correctness and clarity.

11.8. References

[ARS 22] ARSLAN A., COOPER C., KHAN Z. et al., "Artificial intelligence and human workers interaction at team level: A conceptual assessment of the challenges and potential HRM strategies", *International Journal of Manpower*, vol. 43, no. 1, pp. 75–88, doi: 10.1108/IJM-01- 2021-0052, 2022.

[BEE 84] BEER M., SPECTOR B., LAWRENCE P.R. et al., *Managing Human Assets: The Groundwork for General Managers*, Free Press, New York, 1984.

[BUD 22] BUDHWAR P., MALIK A., DE SILVA M.T. et al., "Artificial intelligence – challenges and opportunities for international HRM: A review and research agenda", *The International Journal of Human Resource Management*, vol. 33, no. 6, pp. 1065–1097, doi: 10.1080/09585192.2022.2035161, 2022.

[CAP 17] CAPPELLI P., CONYON M.J., "What do performance appraisals do?", *Industrial and Labor Relations Review*, vol. 71, no. 1, pp. 88–116, doi: 10.1177/0019793917698649, 2017.

[CLA 19] CLAUS L., "HR disruption: Time already to reinvent talent management", *Business Research Quarterly*, vol. 22, no. 3, doi: 10.1016/j.brq.2019.04.002, 2019.

[COR 22] CORON C., "Quantifying human resource management: A literature review", *Personnel Review*, vol. 51, no. 4, pp. 1386–1409, doi: 10.1108/PR-05-2020-0322, 2022.

[DER 16] DEROUS E., DE FRUYT F., "Developments in recruitment and selection research", *International Journal of Selection and Assessment*, vol. 24, no. 1, pp. 1–4, doi: 10.1111/ijsa.12123, 2016.

[DUL 09] DULEBOHN J.H., MOLLOY J.C., PICHLER S.M. et al., "Employee benefits: Literature review and emerging issues", *Human Resource Management Review*, vol. 19, no. 2, pp. 86–103, doi: 10.1016/j.hrmr.2008.10.001, 2009.

[DWI 21] DWIVEDI Y.K., HUGHES L., ISMAGILOVA E. et al., "Artificial Intelligence (AI): Multidisciplinary perspectives on emerging challenges, opportunities, and agenda for research, practice and policy", *International Journal of Information Management*, vol. 57, p. 101994, doi: 10.1016/j.ijinfomgt.2019.08.002, 2021.

[EKU 23] EKUMA K., "Artificial intelligence and automation in human resource development: A systematic review", *Human Resource Development Review*, vol. 23, no. 2, pp. 199–229, doi: 10.1177/15344843231224009, 2023.

[FAR 10] FARNDALE E., SCULLION H., SPARROW P., "The role of the corporate HR function in global talent management", *Journal of World Business*, vol. 45, no. 2, pp. 161–168. doi: 10.1016/j.jwb.2009.09.012, 2010.

[FOM 84] FOMBRUN C.J., TICHY N.M., DEVANNA M.A., *Strategic Human Resource Management*, Wiley, Hoboken, NJ, 1984.

[FRE 04] FREITAG A.R., PICHERIT-DUTHLER G., "Employee benefits communication: Proposing a PRHR cooperative approach", *Public Relations Review*, vol. 30, no. 4, pp. 475–482, doi: 10.1016/j.pubrev.2004.08.006, 2004.

[GOM 22] GOMATHY C.K., RAMASESHACHARYULU A.L.S., SAI SARATH C. et al., "Overview of recruitment and selection process in HRM", *International Journal of Scientific Research in Engineering and Management (IJSREM, 6)*, no. 3, pp. 1–6, doi: 10.55041/IJSREM11714, 2022.

[GÓR 20] GÓRRIZ J., RAMÌREZ J., ORTIZ A. et al., "Artificial intelligence within the interplay between natural and artificial computation: Advances in data science, trends and applications", *Neurocomputing*, vol. 410, pp. 237–270, doi: 10.1016/j.neucom.2020.05.078, 2020.

[GUE 17] GUEST D.E., "Human resource management and employee well-being: Towards a new analytic framework", *Human Resource Management Journal*, vol. 27, no. 1, pp. 22–38, doi: 10.1111/1748-8583.12139, 2017.

[HAE 19] HAENLEIN M., KAPLAN A., "A brief history of artificial intelligence: On the past, present, and future of artificial intelligence", *California Management Review*, vol. 61, no. 4, pp. 5–14, doi: 10.1177/0008125619864925, 2019.

[KAU 15] KAUFMAN B.E., "Evolution of strategic HRM as seen through two founding books: A 30th anniversary perspective on development of the field", *Human Resource Management*, vol. 54, no. 3, pp. 389–407, doi: 10.1002/hrm.21720, 2015.

[KUS 20] KUSTERS R., MISEVIC D., BERRY H. et al., "Interdisciplinary research in artificial intelligence: Challenges and opportunities", *Frontiers in Big Data*, doi: 10.3389/fdata.2020.577974, 2020.

[LIU 18] LIU J., KONG X., XIA F. et al., "Artificial intelligence in the 21st century", *IEEE Access*, vol. 6, pp. 34403–34421, doi: 10.1109/ACCESS.2018.2819688, 2018.

[MAB 03] MABEY C., "Reframing human resource development", *Human Resource Development Review*, vol. 2, no. 4, pp. 430–452, doi: 10.1177/1534484303258042, 2003.

[MAR 23] MARR B., "The difference between generative AI and traditional AI: An easy explanation for anyone", *Forbes*, available at: https://www.forbes.com/sites/bernardmarr/2023/07/24/the-difference-between-generative-aiand-traditional-ai-an-easy-explanation-for-anyone/, 2023.

[MEH 23] MEHAK KUMAR R., MEHTA A., "Artificial intelligence", *International Journal of Advanced Research in Science, Communication and Technology (IJARSCT)*, vol. 3, no. 7, pp. 20–30, doi: 10.48175/IJARSCT-9466, 2023.

[MEY 13] MEYERS C., VAN WOERKOM M., DRIES N., "Talent: Innate or acquired? Theoretical considerations and their implications for talent management", *Human Resource Management Review*, vol. 23, no. 4, pp. 305–321, doi: 10.1016/j.hrmr.2013.05.003,2013.

[MOR 22] MORANDÍN-AHUERMA F., "What is artificial intelligence?", *International Journal of Research Publication and Reviews*, vol. 3, no. 12, pp. 1947–1951, doi: 10.55248/gengpi.2022.31261, 2022.

[PAN 22] PAN Y., FROESE F., LIU N. et al., "The adoption of artificial intelligence in employee recruitment: The influence of contextual factors", *The International Journal of Human Resource Management*, vol. 33, no. 6, pp. 1125–1147, doi: 10.1080/09585192.2021.1879206, 2022.

[POD 20] PODGORODNICHENKO N., EDGAR F., MCANDREW I., "The role of HRM in developing sustainable organizations: Contemporary challenges and contradictions", *Human Resource Management Review*, vol. 30, no. 3, p. 100685, doi: 10.1016/j.hrmr.2019.04.001, 2020.

[POT 21] POTOČNIK K., ANDERSON N.R., BORN M. et al., "Paving the way for research in recruitment and selection: Recent developments, challenges and future opportunities", *European Journal of Work and Organizational Psychology*, vol. 30, no. 2, pp. 159–174, doi: 10.1080/1359432X.2021.1904898, 2021.

[QAM 21] QAMAR Y., AGRAWAL R.K., SAMAD T.A., "When technology meets people: The interplay of artificial intelligence and human resource management", *Journal of Enterprise Information Management*, vol. 34, no. 5, pp. 1339–1370, doi: 10.1108/JEIM-11-2020-0436, 2021.

[REI 08] REILLY P., "Identifying the right course for talent management", *Public Personnel Management*, vol. 37, no. 4, pp. 381–392, doi: 10.1177/009102600803700401, 2008.

[ROB 03] ROBERTS G.E., "Employee performance appraisal system participation: A technique that works", *Public Personnel Management*, vol. 32, no. 1, pp. 89–98, doi: 10.1177/009102600303200105, 2003.

[SAE 23] SAEED W., OMLIN C., "Explainable AI (XAI): A systematic meta-survey of current challenges and future opportunities", *Artificial Intelligence Review*, doi: 10.1016/j.knosys. 2023.110273, 2023.

[SAN 03] SANTOS A., STUART M., "Employee perceptions and their influence on training effectiveness", *Human Resource Management Journal*, vol. 13, no. 1, pp. 27–45, doi: 10.1111/j.1748- 8583.2003.tb00082.x, 2003.

[SHA 17] SHARMA A., SHARMA T., "HR analytics and performance appraisal system: A conceptual framework for employee performance improvement", *Management Research Review*, vol. 40, no. 6, pp. 684–697, doi: 10.1108/MRR-04-2016-0084, 2017.

[SOD 06] SODERLUND J., BREDIN K., "HRM in project-intensive firms: Changes and challenges", *Human Resource Management*, vol. 45, no. 2, pp. 249–265, doi: 10.1002/ hrm.20107, 2006.

[STO 15] STONE D.L., DEADRICK D.L., "Challenges and opportunities affecting the future of human resource management", *Human Resource Management Review*, vol. 25, no. 2, pp. 139–145, doi: 10.1016/j.hrmr.2015.01.003, 2015.

[SUE 19] SUEN H.Y., CHEN M.Y.C., LU S.H., "Does the use of synchrony and artificial intelligence in video interviews affect interview ratings and applicant attitudes?", *Computers in Human Behavior*, vol. 98, pp. 93–101, doi: 10.1016/j.chb.2019.04.012, 2019.

[TAM 19] TAMBE P., CAPPELLI P., YAKUBOVICH V., "Artificial intelligence in human resources management: Challenges and a path forward", *California Management Review*, vol. 61, no. 4, pp. 15–42, 2019.

[TAR 10] TARIQUE I., SCHULER R., "Global talent management: Literature review, integrative framework, and suggestions for further research", *Journal of World Business*, vol. 45, no. 2, pp. 122–133, doi: 10.1177/0008125619867910, 2010.

[THO 11] THOMPSON P., "The trouble with HRM", *Human Resource Management Journal*, vol. 21, no. 4, pp. 355–367, doi: 10.1111/j.1748-8583.2011.00180.x, 2011.

[TOM 20] TOMAŠEV, N., CORNEBISE, J., HUTTER, F. et al., "AI for social good: Unlocking the opportunity for positive impact", *Nature Communications*, vol. 11, no. 1, p. 2468, doi: 10.1038/s41467-020-15871-z, 2020.

[URB 21] URBANCOV H., VRABCOV P., HUD KOV M. et al., "Effective training evaluation: The role of factors influencing the evaluation of effectiveness of employee training and development", *Sustainability*, vol. 13, no. 5, p. 2721, doi: 10.3390/su13052721, 2021.

[WAN 19] WANG P., "On defining artificial intelligence", *Journal of Artificial General Intelligence*, vol. 10, no. 2, pp. 1–37, doi: 10.2478/jagi-2019-0002, 2019.

[WOH 24] WOHLFARTH A., "Gen AI vs. Traditional AI: Understanding the differences. *OurCrowd*, available at: https://www.ourcrowd.com/learn/generative-ai-vs-traditional-ai, 2024.

[WRI 02] WRIGHT P.M., BOSWELL W.R., Desegregating HRM: A review and synthesis of micro and macro human resource management research, CAHRS Working Paper Series, Cornell University, available at: https://hdl.handle.net/1813/77382, 2002.

List of Authors

Martha Elena Moreno BARBOSA
Tecnológico de Monterrey
Jal
Mexico

Patrick BARETTO
Weiden Business School
Ostbayerische Technische Hochschule (OTH)
Weiden
Germany

Pedro Miguel Torres BARROS
School of Economics, Management and Political Sciences
University of Minho
Braga
Portugal

Laura Esther Zapata CANTÚ
Tecnológico de Monterrey
Nuevo Leon
Mexico

J. Paulo DAVIM
Department of Mechanical Engineering
University of Aveiro
Campus de Santiago
Aveiro
Portugal

Feri FERDIAN
Faculty of Tourism and Hospitality
Universitas Negeri Padang
Padang
Indonesia

Yudha Aditya FIANDRA
Faculty of Engineering
Universitas Negeri Padang
Padang
Indonesia

Florin Sabin FOLTEAN
Faculty of Economics and Business Administration
West University of Timisoara
Timisoara
Romania

Octavian Dumitru HERA
Faculty of Economics and Business Administration
West University of Timisoara
Timisoara
Romania

Qeis KAMRAN
Weiden Business School
Ostbayerische Technische Hochschule (OTH)
Weiden
Germany

Cristiana Cerqueira LEAL
Centre for Research in Economics and Management (NIPE)
and
School of Economics, Management and Political Science
University of Minho
Braga
Portugal

Ana Valdés LOYOLA
Universidad de Monterrey
Nuevo Leon
Mexico

Carolina MACHADO
Department of Management
School of Economics, Management and Political Science
and
Interdisciplinary Centre of Social Sciences (CICS.NOVA.UMinho)
University of Minho
Braga
Portugal

Mitra MADANCHIAN
School of Arts, Science and Technology
University Canada West
Vancouver
Canada

Xabier MARTÍNEZ-ROLÁN
Universidade de Vigo
Pontevedra
Spain

Daniel Javier de la Garza MONTEMAYOR
Universidad de Monterrey
Nuevo Leon
Mexico

Flor MORTON
Department of Management
Universidad de Monterrey
Nuevo Leon
Mexico

Benilde OLIVEIRA
School of Economics, Management and Political Science
University of Minho
Braga
Portugal

Soha RAWAS
Faculty of Science
Department of Mathematics and Computer Science
Beirut Arab University
Beirut
Lebanon

Agariadne Dwinggo SAMALA
Faculty of Engineering
Universitas Negeri Padang
Padang
Indonesia

Jesús Pérez SEOANE
Universidade de Vigo
Pontevedra
Spain

Cerine TAFRAN
Faculty of Science
Department of Mathematics and Computer Science
Beirut Arab University
Beirut
Lebanon

Hamed TAHERDOOST
School of Arts, Science and Technology
University Canada West
Vancouver
Canada

Teresa TREVIÑO-BENAVIDES
Department of Management
Universidad de Monterrey
Nuevo Leon
Mexico

Juan Manuel Corbacho VALENCIA
Universidade de Vigo
Pontevedra
Spain

Index

A, B

C, D

E, F

G, H

I, K

L, M

N, O

P, R

S, T

U, V, W

Other titles from

ISTE

in

Computer Engineering

2026

BHAMBRI Pankaj, DUTTA Pushan Kumar, KHAN Mudassir, STAROSTKA-PATYK Marta
Next-Gen Supply Chains: AI, Automation, and Sustainability in a Disrupted World

DARCHE Philippe
Computer Memories 1: Core Concepts
Computer Memories 2: Static Random-Access Memory

MAHOUT Vincent, ROCACHER Thierry, AURIOL Guillaume
Programming of Embedded Systems: Applications to SM32F10x µControllers

2025

CHIKH Azeddine
Systemic Approach to Categorizing and Modeling Requirements

KUMAR Abhishek, BATTA Priya, AHUJA Sachin, RATHORE Pramod Singh
AI-driven Innovations in Physiotherapy and Oncology 1
AI-driven Innovations in Physiotherapy and Oncology 2

TANEJA Ashu, KUMAR Abhishek, LIMKAR Suresh Vishnudas, OUAISSA Mariya, OUAISSA Mariyam
Driving Innovation through AI and Digital Twin for 6G Powered Sustainable Ultra Smart Cities

TANEJA Ashu, KUMAR Abhishek, LIMKAR Suresh Vishnudas, OUAISSA Mariya, OUAISSA Mariyam
6G Urban Innovation: AI and Digital Twin for Next-Gen Sustainable Cities

2024

CHAKRABORTY Rajdeep, GHOSH Anupam, KUMAR MANDAL Jyotsna, CHOUDHURY Tanupriya, CHATTERJEE Prasenjit
Smart Edge Computing: An Operation Research Perspective
(International Perspectives in Decision Analytics and Operations Research Set – Volume 2)

HOMÈS Bernard
Fundamentals of Software Testing (2nd edition, revised and updated)

JERBI Abdessalem
Discrete-Event Simulation: Concepts and Production in Arena

PRADHAN Devasis, KUMAR SAHU Prasanna, MYO TUN Hla, CHATTERJEE Prasenjit
Artificial and Cognitive Computing for Sustainable Healthcare Systems in Smart Cities
(International Perspectives in Decision Analytics and Operations Research Set – Volume 3)

2023

CHOUDHURY Tanupriya, KHANNA Abhirup, CHATTERJEE Prasenjit, UM Jung-Sup, BHATTACHARYA Abhishek
Blockchain Applications in Healthcare: Innovations and Practices
(International Perspectives in Decision Analytics and Operations Research Set – Volume 1)

PAI G A Vijayalakshmi
A Textbook of Data Structures and Algorithms 1: Mastering Linear Data Structures
A Textbook of Data Structures and Algorithms 2: Mastering Nonlinear Data Structures
A Textbook of Data Structures and Algorithms 3: Mastering Advanced Data Structures and Algorithm Design Strategies

2022

HOMÈS Bernard
Advanced Testing of Systems-of-Systems 1: Theoretical Aspects
Advanced Testing of Systems-of-Systems 2: Practical Aspects

MEHTA Shikha, TIWARI Sanju, SIARRY Patrick, JABBAR M.A.
Tools, Languages, Methodologies for Representing Semantics on the Web of Things

SIDHOM Sahbi, KADDOUR Amira
Systems and Uses of Digital Sciences for Knowledge Organization (Digital Tools and Uses Set – Volume 9)

ZAIDOUN Ameur Salem
Computer Science Security: Concepts and Tools

2021

DARCHE Philippe
Microprocessor 4: Core Concepts – Software Aspects
Microprocessor 5: Software and Hardware Aspects of Development, Debugging and Testing – The Microcomputer

DELHAYE Jean-Loïc
Inside the World of Computing: Technologies, Uses, Challenges

HARDIN Thérèse, JAUME Mathieu, PESSAUX François,
VIGUIÉ DONZEAU-GOUGE Véronique
Concepts and Semantics of Programming Languages 1: A Semantical Approach with OCaml and Python
Concepts and Semantics of Programming Languages 2: Modular and Object-oriented Constructs with OCaml, Python, C++, Ada and Java

MKADMI Abderrazak
Archives in The Digital Age: Preservation and the Right to be Forgotten (Digital Tools and Uses Set – Volume 8)

TOKLU Yusuf Cengiz, BEKDAŞ Gebrail, NIGDELI Sinan Melih
Metaheuristics for Structural Design and Analysis (Optimization Heuristics Set – Volume 3)

2020

DARCHE Philippe
Microprocessor 1: Prolegomenes – Calculation and Storage Functions – Models of Computation and Computer Architecture
Microprocessor 2: Core Concepts – Communication in a Digital System
Microprocessor 3: Core Concepts – Hardware Aspects

DUVAUT Patrick, DALLOZ Xavier, MENGA David, KOEHL François,
CHRIQUI Vidal, BRILL Joerg
Internet of Augmented Me, I.AM: Empowering Innovation for a New Sustainable Future

LAFFLY Dominique
TORUS 1 – Toward an Open Resource Using Services: Cloud Computing for Environmental Data
TORUS 2 – Toward an Open Resource Using Services: Cloud Computing for Environmental Data
TORUS 3 – Toward an Open Resource Using Services: Cloud Computing for Environmental Data

LAURENT Anne, LAURENT Dominique, MADERA Cédrine
Data Lakes (Databases and Big Data Set – Volume 2)

OULHADJ Hamouche, DAACHI Boubaker, MENASRI Riad
Metaheuristics for Robotics
(Optimization Heuristics Set – Volume 2)

SADIQUI Ali
Computer Network Security

VENTRE Daniel
Artificial Intelligence, Cybersecurity and Cyber Defense

2019

BESBES Walid, DHOUIB Diala, WASSAN Niaz, MARREKCHI Emna
Solving Transport Problems: Towards Green Logistics

CLERC Maurice
Iterative Optimizers: Difficulty Measures and Benchmarks

GHLALA Riadh
Analytic SQL in SQL Server 2014/2016

TOUNSI Wiem
Cyber-Vigilance and Digital Trust: Cyber Security in the Era of Cloud Computing and IoT

2018

ANDRO Mathieu
Digital Libraries and Crowdsourcing
(Digital Tools and Uses Set – Volume 5)

ARNALDI Bruno, GUITTON Pascal, MOREAU Guillaume
Virtual Reality and Augmented Reality: Myths and Realities

CARDON Alain
Beyond Artificial Intelligence: From Human Consciousness to Artificial Consciousness

HOMAYOUNI S. Mahdi, FONTES Dalila B.M.M.
Metaheuristics for Maritime Operations
(Optimization Heuristics Set – Volume 1)

JEANSOULIN Robert
JavaScript and Open Data

PIVERT Olivier
NoSQL Data Models: Trends and Challenges
(Databases and Big Data Set – Volume 1)

SEDKAOUI Soraya
Data Analytics and Big Data

SALEH Imad, AMMI Mehdi, SZONIECKY Samuel
Challenges of the Internet of Things: Technology, Use, Ethics
(Digital Tools and Uses Set – Volume 7)

SZONIECKY Samuel
Ecosystems Knowledge: Modeling and Analysis Method for Information and Communication
(Digital Tools and Uses Set – Volume 6)

2017

BENMAMMAR Badr
Concurrent, Real-Time and Distributed Programming in Java: Threads, RTSJ and RMI

BOUHAÏ Nasreddine, SALEH Imad
Internet of Things: Evolutions and Innovations
(Digital Tools and Uses Set – Volume 4)

HÉLIODORE Frédéric, NAKIB Amir, ISMAIL Boussaad, OUCHRAA Salma, SCHMITT Laurent
Metaheuristics for Intelligent Electrical Networks
(Metaheuristics Set – Volume 10)

MA Haiping, SIMON Dan
Evolutionary Computation with Biogeography-based Optimization
(Metaheuristics Set – Volume 8)

PÉTROWSKI Alain, BEN-HAMIDA Sana
Evolutionary Algorithms
(Metaheuristics Set – Volume 9)

PAI G A Vijayalakshmi
Metaheuristics for Portfolio Optimization: An Introduction using MATLAB®
(Metaheuristics Set – Volume 11)

REYES-GARCIA Everardo
The Image-Interface: Graphical Supports for Visual Information
(Digital Tools and Uses Set – Volume 3)

REYES-GARCIA Everardo, BOUHAÏ Nasreddine
Designing Interactive Hypermedia Systems
(Digital Tools and Uses Set – Volume 2)

SZONIECKY Samuel, BOUHAÏ Nasreddine
Collective Intelligence and Digital Archives: Towards Knowledge Ecosystems
(Digital Tools and Uses Set – Volume 1)

2016

BLUM Christian, FESTA Paola
Metaheuristics for String Problems in Bio-informatics
(Metaheuristics Set – Volume 6)

DEROUSSI Laurent
Metaheuristics for Logistics
(Metaheuristics Set – Volume 4)

DHAENENS Clarisse, JOURDAN Laetitia
Metaheuristics for Big Data
(Metaheuristics Set – Volume 5)

LABADIE Nacima, PRINS Christian, PRODHON Caroline
Metaheuristics for Vehicle Routing Problems
(Metaheuristics Set – Volume 3)

LEROY Laure
Eyestrain Reduction in Stereoscopy

LUTTON Evelyne, PERROT Nathalie, TONDA Alberto
Evolutionary Algorithms for Food Science and Technology (Metaheuristics Set – Volume 7)

MAGOULÈS Frédéric, ZHAO Hai-Xiang
Data Mining and Machine Learning in Building Energy Analysis

RIGO Michel
Advanced Graph Theory and Combinatorics

ZELM Martin, DOUMEINGTS Guy, MENDONÇA Joao Pedro
Enterprise Interoperability: in the Digitized and Networked Factory of the Future

2015

BARBIER Franck, RECOUSSINE Jean-Luc
COBOL Software Modernization: From Principles to Implementation with the BLU AGE® Method

CHEN Ken
Performance Evaluation by Simulation and Analysis with Applications to Computer Networks

CLERC Maurice
Guided Randomness in Optimization (Metaheuristics Set – Volume 1)

DURAND Nicolas, GIANAZZA David, GOTTELAND Jean-Baptiste, ALLIOT Jean-Marc
Metaheuristics for Air Traffic Management (Metaheuristics Set – Volume 2)

MAGOULÈS Frédéric, ROUX François-Xavier, HOUZEAUX Guillaume
Parallel Scientific Computing

MUNEESAWANG Paisarn, YAMMEN Suchart
Visual Inspection Technology in the Hard Disk Drive Industry

2014

BOULANGER Jean-Louis
Formal Methods Applied to Industrial Complex Systems
Formal Methods Applied to Complex Systems: Implementation of the B Method

EXPOSITO Ernesto, DIOP Codé
Smart SOA Platforms in Cloud Computing Architectures

GARDI Frédéric, BENOIST Thierry, DARLAY Julien, ESTELLON Bertrand, MEGEL Romain
Mathematical Programming Solver based on Local Search

KRICHEN Saoussen, CHAOUACHI Jouhaina
Graph-related Optimization and Decision Support Systems

LARRIEU Nicolas, VARET Antoine
Rapid Prototyping of Software for Avionics Systems: Model-oriented Approaches for Complex Systems Certification

LAURAS Matthieu, ZELM Martin, ARCHIMÈDE Bernard, BÉNABEN Frédérick, DOUMEINGTS Guy
Enterprise Interoperability: Interoperability for Agility, Resilience and Plasticity of Collaborations

OUSSALAH Mourad Chabane
Software Architecture 1
Software Architecture 2

QUESNEL Flavien
Scheduling of Large-scale Virtualized Infrastructures: Toward Cooperative Management

RIGO Michel
Formal Languages, Automata and Numeration Systems 1: Introduction to Combinatorics on Words
Formal Languages, Automata and Numeration Systems 2: Applications to Recognizability and Decidability

SAINT-DIZIER Patrick
Musical Rhetoric: Foundations and Annotation Schemes

TOUATI Sid, DUPONT DE DINECHIN Benoit
Advanced Backend Code Optimization

2013

ANDRÉ Étienne, SOULAT Romain
The Inverse Method: Parametric Verification of Real-time Embedded Systems

BOULANGER Jean-Louis
Safety Management for Software-based Equipment

DELAHAYE Daniel, PUECHMOREL Stéphane
Modeling and Optimization of Air Traffic

FRANCOPOULO Gil
LMF — Lexical Markup Framework

GHEDIRA Khaled
Constraint Satisfaction Problems: CSP Formalisms and Techniques

GOVAERT Gérard, NADIF Mohamed
Co-Clustering: Models, Algorithms and Applications

LANDRAGIN Frédérick
Man–Machine Dialogue: Design and Challenges

LUCAS Laurent, LOSCOS Céline, REMION Yannick
3D Video: Models, From Capture to Diffusion

ROCHANGE Christine, UHRIG Sascha, SAINRAT Pascal
Time-Predictable Architectures

WAHBI Mohamed
Algorithms and Ordering Heuristics for Distributed Constraint Satisfaction Problems

ZELM Martin, VAN SINDEREN Marten, FERRAIRA PIRES Luis, DOUMEINGTS Guy
Enterprise Interoperability: Research and Applications in the Service-oriented Ecosystem

2012

ARBOLEDA Hugo, ROYER Jean-Claude
Model-Driven and Software Product Line Engineering

BLANCHET Gérard, DUPOUY Bertrand
Computer Architecture

BOULANGER Jean-Louis
Industrial Use of Formal Methods: Formal Verification
Formal Methods: Industrial Use from Model to the Code

CALVARY Gaëlle, DELOT Thierry, SÈDES Florence, TIGLI Jean-Yves
Computer Science and Ambient Intelligence

MAHOUT Vincent
Assembly Language Programming: ARM Cortex-M3 2.0: Organization, Innovation and Territory

MARLET Renaud
Program Specialization

SOTO Maria, SEVAUX Marc, ROSSI André, LAURENT Johann
Memory Allocation Problems in Embedded Systems: Optimization Methods

2011

BICHOT Charles-Edmond, SIARRY Patrick
Graph Partitioning

BOULANGER Jean-Louis
Static Analysis of Software: The Abstract Interpretation

CAFERRA Ricardo
Logic for Computer Science and Artificial Intelligence

KORDON Fabrice, HADDAD Serge, PAUTET Laurent, PETRUCCI Laure
Distributed Systems: Design and Algorithms
Models and Analysis in Distributed Systems

LORCA Xavier
Tree-based Graph Partitioning Constraint

RIDHA MAHJOUB A.
Progress in Combinatorial Optimization: ISCO2010

TRUCHET Charlotte, ASSAYAG Gerard
Constraint Programming in Music

VICAT-BLANC Pascale, GOGLIN Brice, GUILLIER Romaric,
SOUDAN Sébastien
Computing Networks: From Cluster to Cloud Computing

2010

AUBRUN Christophe, SIMON Daniel, SONG Ye-Qiong
Co-design Approaches for Dependable Networked Control Systems

AUDIBERT Pierre
Mathematics for Informatics and Computer Science

BABAU Jean-Philippe, BLAY-FORNARINO Mireille, CHAMPEAU Joël,
ROBERT Sylvain, SABETTA Antonino
Model Driven Engineering for Distributed Real-Time Systems: MARTE Modeling, Model Transformations and their Usages

BOULANGER Jean-Louis
Safety of Computer Architectures

SIGAUD Olivier, BUFFET Olivier
Markov Decision Processes in Artificial Intelligence

MONMARCHÉ Nicolas, GUINAND Frédérick, SIARRY Patrick
Artificial Ants

SOLNON Christine
Ant Colony Optimization and Constraint Programming

2009

FOURNIER Jean-Claude
Graph Theory and Applications: with exercises and problems

GUÉDON Jeanpierre
The Mojette Transform: theory and applications

JARD Claude, ROUX Olivier H.
Communicating Embedded Systems: Software and Design

LECOUTRE Christophe
Constraint Networks: Techniques and Algorithms

2008

BANÂTRE Michel, MARRON Pedro Jose, OLLERO Anibal, WOLITZ Adam
Cooperating Embedded Systems and Wireless Sensor Networks

MERZ Stephan, NAVET Nicolas
Modeling and Verification of Real-time Systems: Formalisms and Software Tools

PASCHOS Vangelis Th.
Combinatorial Optimization and Theoretical Computer Science: Interfaces and Perspectives

WALDNER Jean-Baptiste
Nanocomputers and Swarm Intelligence

2007

BENHAMOU Frédéric, JUSSIEN Narendra, O'SULLIVAN Barry
Trends in Constraint Programming

JUSSIEN Narendra
A to Z of Sudoku

2006

BABAU Jean-Philippe, CHAMPEAU Joël, GÉRARD Sébastien
From MDD Concepts to Experiments and Illustrations – DRES 2006

HABRIAS Henri, FRAPPIER Marc
Software Specification Methods: An Overview Using a Case Study

MURAT Cécile, PASCHOS Vangelis Th.
Probabilistic Combinatorial Optimization on Graphs

PANETTO Hervé, BOUDJLIDA Nacer
Interoperability for Enterprise Software and Applications: Proceedings of the Workshops and the Doctorial Symposium of the Second IFAC/IFIP I-ESA International Conference: EI2N, WSI, IS-TSPQ 2006

2005

GÉRARD Sébastien, BABAU Jean-Philippe, CHAMPEAU Joël
Model Driven Engineering for Distributed Real Time Embedded Systems

PANETTO Hervé
Interoperability of Enterprise Software and Applications: Workshops of the INTEROP-ESA 2005 International Conference (EI2N, WSI, ISIDI and IEHENA 2005)

Printed and bound by CPI Group (UK) Ltd, Croydon, CR0 4YY
13/09/2026
14961013-0007